Photovoltaic Systems

Yaman Abou Jieb • Eklas Hossain

Photovoltaic Systems

Fundamentals and Applications

 Springer

Yaman Abou Jieb
Wilsonville
OR, USA

Eklas Hossain
Oregon Institute of Technology
Klamath Falls
OR, USA

ISBN 978-3-030-89779-6   ISBN 978-3-030-89780-2   (eBook)
https://doi.org/10.1007/978-3-030-89780-2

This Springer imprint is published by the registered company Springer Nature Switzerland AG
The registered company address is: Gewerbestrasse 11, 6330 Cham, Switzerland

Preface

The increasing deployment of solar photovoltaic systems across the globe and their promising future has ushered a new era in the world of renewable energy technologies. As the world seeks cleaner ways of generating energy and opts out of using fossil fuels, solar photovoltaic systems are predicted to be the dominant source of energy within a few decades. According to the Photovoltaic Market Outlook— 2026 by Allied Market Research, the global photovoltaic market is worth $53,916 million as of 2018 and is expected to reach $333,725 million within 2026. Therefore, this technology is highly promising to supply a major fraction of the world's energy demands. So, it is essential for everyone to have a basic understanding of the photovoltaic technology and its applications in various fields.

Solar photovoltaic systems are commonly deployed all over the world in residential, commercial, and industrial buildings as rooftop, pole-mounted, or ground mounted installations. They are ubiquitous and have diverse applications. Due to the omnipresence of solar photovoltaic systems, the authors felt prompted to write a book on photovoltaic systems, focusing on the fundamentals and applications. This book is a guideline on designing practical solar photovoltaic systems. The authors ensured that simplicity is maintained throughout the book, so that it aids everyone irrespective of their backgrounds. Engineering students can use the book as a reference for courses on photovoltaic systems, renewable energy, or electric power systems. They can also use the book for designing projects and conducting research and experimentations. Teachers may also use the book as a lecture material and prepare their lecture outlines based on this book. Professional engineers and designers will also find this book generous in examples and practical insights that will help to design cost-effective photovoltaic systems.

The navy blue or black flat solar panels are more than just light-absorbing devices. They convert the light energy of the Sun into usable electricity that powers our energy needs. A solar array consists of several solar modules, and each module consists of multiple solar cells. For a solar photovoltaic system, in addition to the solar array, several other components are necessary to build a completely functioning system. To understand how a solar photovoltaic system works, a basic understanding of the physics behind the working principle of the solar cell is required, together with a fairly good knowledge about the other components such as battery, charge controller, inverter, wiring, protection devices, and so on. All these topics are covered in this book. This book is rich in its step-by-step description of the design of grid-tied, off-grid, hybrid, and distributed photovoltaic systems.

This book is a fundamental guide to the solar photovoltaic technology and is a designer's handbook for installing photovoltaic systems. With step-by-step instructions, calculations, and illustrations, this book will equip the reader with all the jargons and provide a first-hand knowledge on designing photovoltaic systems. The book is designed to provide a hands-on approach to learning about solar photovoltaic systems for students and professionals alike. The first three chapters introduce and discuss the fundamentals of solar photovoltaic systems, the fourth chapter explains the assessment of solar resources, the fifth chapter describes the components used in a solar photovoltaic system, and finally, the sixth and the seventh chapters demonstrate the detailed design procedure and case studies of practical solar photovoltaic system installations.

Key Features of This Book

- Reader-friendly, clear, and concise description of all core topics of solar photovoltaic technology
- Practical case studies to help understand the concepts from a realistic approach
- Applications of solar photovoltaics in different purposes
- Elaboration of related concepts such as battery energy storage, inverters, charge controllers, grid integration, etc.
- Vivid illustrations and real images of photovoltaic systems

Wilsonville, OR, USA   Yaman Abou Jieb

Klamath Falls, OR, USA   Eklas Hossain
September 2021

Acknowledgments

The authors would like to solemnly appreciate the contribution of everyone at the Springer Nature who were involved with the production of this book. We are also thankful to the authors of innumerable research papers, books, and other internet resources that were sought for gaining the knowledge to write this book. We are immensely grateful to the kind reviewers of this book, whose attentive scrutiny has enabled us to get rid of many flaws and deliver an improved version of our initial draft. Their selfless and voluntary contribution truly deserves a heartfelt gratitude and a deep appreciation. We acknowledge the Oregon Institute of Technology and its contribution in empowering renewable energy technologies, particularly solar photovoltaic systems. Many real case studies and real images have been utilized in this book to aid comprehension and visual learning. We thank the owners of those photovoltaic systems for allowing us to use images of the systems for this book. In addition, we thank the creators of the software MATLAB, PVsyst, Homer Pro, and Dialux for their amazing creations that have allowed us to present relatable and realistic information in this book. We are also immensely grateful to our families, for supporting us throughout our lives. Lastly, we are indebted to our Creator, who has blessed us with beautiful lives and the knowledge for writing this book.

Contents

About the Authors

Yaman Abou Jieb is an electrical power engineer with a master's degree in renewable energy engineering from the Oregon Institute of Technology (OIT), which is home to the only ABET-accredited BS and MS programs in renewable energy engineering. During his master's degree studies, he was the instructor of electrical circuits sequence and the teaching assistant of the power and power electronics sequences for graduate students.

Prior to joining OIT, Yaman started his own company in the Middle East that provides sustainable energy solutions by designing and installing all possible PV off-grid applications such as water pumping, street lighting, telecommunication sites, and residential applications. Moreover, he designed commercial and utility-scale grid-tied systems and provided technical due diligence in servicing those plants. After receiving his master's degree, Yaman occupied the position of senior service engineer at the SMA America where he was providing high-level technical support for difficult and unresolved customer inquiries for utility-scale solar and energy storage inverters in North and South America. Yaman is currently an electrical power engineer at the Atwell, LLC, where he provides electrical supervision of utility-scale solar PV and battery storage design projects in the USA.

Eklas Hossain (M'09, SM'17) received his PhD from the College of Engineering and Applied Science at University of Wisconsin Milwaukee (UWM). He received his MS in Mechatronics and Robotics Engineering from the International Islamic University of Malaysia, Malaysia, in 2010 and BS in Electrical and Electronic Engineering from the Khulna University of Engineering and Technology, Bangladesh, in 2006.

The author has been working in the area of distributed power systems and renewable energy integration for last 10 years, and he has published a number of research papers and posters in this field. He is now involved with several research projects on renewable energy and grid-tied microgrid system at the Oregon Tech, as an Associate Professor in the Department of Electrical Engineering and Renewable Energy and as a Senior Electrical Consultant at the RRC Power and Energy. He is the senior member of Association of Energy Engineers (AEE). He is currently serving as an Associate Editor of IEEE Access. He is working as an Associate Researcher at the Oregon Renewable Energy Center (OREC). He is a registered Professional

Engineer (PE) in the state of Oregon, USA. He is also a Certified Energy Manager (CEM) and a Renewable Energy Professional (REP).

His research interests include modeling, analysis, design, and control of power electronic devices; energy storage systems; renewable energy sources; integration of distributed generation systems; microgrid and smart grid applications; robotics, and advanced control system.

Dr. Hossain has authored the book *"Excel Crash Course for Engineers,"* co-authored the book *"Renewable Energy Crash Course - A Concise Introduction,"* and is working on several other book projects. He is the winner of the Rising Faculty Scholar Award in 2019 and the Faculty Achievement Award in 2020 from the Oregon Institute of Technology for his outstanding contribution in research and teaching. Dr. Hossain, with his dedicated research team, is looking forward to exploring methods to make electric power systems more sustainable, cost-effective, and secure through extensive research and analysis on energy storage, microgrid system, and renewable energy sources.

Solar Photovoltaic Industry Overview

1.1 The Emergence of Renewable Energy Resources

The conventional energy sources are fossil fuels such as natural gas, oil, and coal, which have been the dominant energy resources in the global electricity generation industry since the industrial revolution. However, no one had seriously thought about replacing them with other resources due to their abundance and low prices. The oil crisis was first perceived in 1973 by the Organization of Petroleum Exporting Countries (OPEC) to decrease the production and increase the price. The direct result of this crisis was a drastic increase in oil prices by 400% (from $3/barrel up to $12/barrel). Moreover, it showed the world the necessity of finding a nondepletable and local energy source. In addition, the emissions from the combustion of fossil fuels and the resulting detrimental impacts on the environment and the climate have urged global minds to think about using cleaner energy sources. Since then, the renewable energy (RE) industry started to get governmental funds and support, especially solar energy, which is considered one of the two most propitious RE resources besides wind energy. The other popular forms of RE are hydroelectric power, geothermal energy, bioenergy, and wave and tidal energy.

1.2 Solar Energy Technologies

The energy coming to the Earth from the Sun is known as solar energy. Solar energy is a vast and free source that can easily meet the global energy demand. The amount of energy in sunlight is tantamount to 10,000 times more than the global energy demand. Therefore, the potential of solar energy is unparalleled to fulfill the energy requirements of all life forms on our planet. Three main technologies are used for harvesting energy from the Sun, namely solar heating and cooling (SHC), concentrated solar power (CSP), and solar photovoltaics (PV) [1]. Solar irradiation consists mostly of heat energy (infrared (IR) radiation), followed by light energy

© The Author(s), under exclusive license to Springer Nature Switzerland AG 2022
Y. Abou Jieb, E. Hossain, *Photovoltaic Systems*,
https://doi.org/10.1007/978-3-030-89780-2_1

and ultraviolet (UV) radiation. All parts of the solar spectrum can be tapped into to derive usable energy. The following three technologies utilize different parts of the solar spectrum to produce usable energy.

1.2.1 Solar Heating and Cooling Technology

The SHC technology is based on concentrating sunlight to use thermal energy from the Sun to heat water. It is widely used in residential applications, in large-scale commercial projects such as swimming pools, and in industries that use hot water in the production process, such as in the food and chemical industries. Using conventional mechanical principles, users can utilize hot water for space heating or cooling purposes by replacing traditional electrical heating elements or fuel-based heaters with flat or tubular solar collectors. Solar water heaters are efficient and economically feasible. Therefore, some countries such as Spain mandated all new homes to install solar water heaters on their roofs to eliminate the usage of high energy-consuming electrical heaters [2]. SHC technology is classified into three main categories:

1. Solar water heating system (SWHS)
2. Heating, ventilation, and cooling (HVAC)
3. Solar industrial heating process (SIHP)

1.2.2 Concentrated Solar Power Technology

The concentrated solar power technology or CSP uses solar irradiance directly to generate steam that is injected into a steam turbine to convert the mechanical energy into electrical energy. Flat, dish-shaped, or parabolic mirrors are used to reflect solar rays and concentrate them on a receiver tube or container that contains a heat transfer fluid (HTF) to produce high-pressure steam, which is then released to a conventional turbine. All reflectors are automatically controlled to track the Sun path to receive and reflect the solar irradiance appropriately. CSP technologies can be concentrating or non-concentrating, which means that either they can concentrate the sunbeams and then utilize them or they can directly use the sunlight. The complete classification of CSP can be roughly represented as:

1. Parabolic trough collector
2. Parabolic dish collector
3. Linear Fresnel collector
4. Solar power tower

The parabolic trough collector is the most developed CSP technology at present. But the popularity of the solar power tower is also precipitously rising, as several mega-scale solar projects are employing this technology. The schematic of the solar

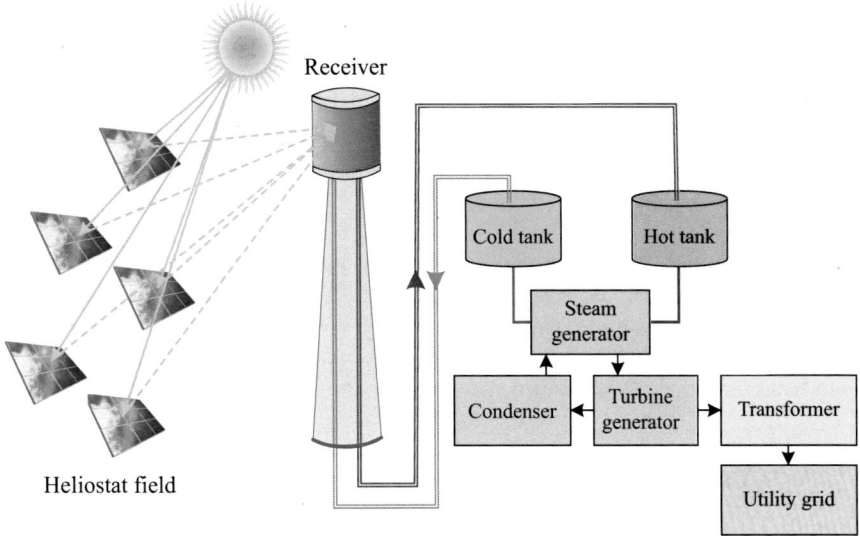

Receiver

Cold tank   Hot tank

Steam generator

Condenser ← Turbine generator → Transformer

Heliostat field

Utility grid

Fig. 1.1 Schematic of the solar power tower technology

power technology is depicted in Fig. 1.1. This technology involves a tall tower surrounded by heliostat mirrors or reflectors installed spanning across a large area. Sunlight falls on these mirrors and is reflected toward the top of the tower. The positioning and orientation of the heliostat mirrors are adjusted accordingly with Sun trackers to ensure that most of the sunlight is directed straight to the tower head. Inside the tower head is a heat transfer fluid (HTF), which absorbs the heat from the sunlight. The HTF is then transferred into a hot tank, where the heat is conducted into a steam generator to generate steam using the heat. The steam is then used to energize a conventional turbine to drive a generator to generate electricity. Finally, through a transformer, the power is fed to the utility grid. Meanwhile, the HTF is cooled in a cold tank and pumped up the tower to repeat the cycle. The HTF is usually steam or molten salt, but the latter is preferred due to its superior heat capacity and transfer abilities.

To know more about the vast spectra of CSP technologies, have a look at Ref. [3].

1.2.3   Solar PV Technology

The word photovoltaic implies the conversion of "photo" or light into "volts" or electricity. The PV technology is the only solar technology that directly converts the sunlight into electricity in direct current (DC) form using semiconductor materials fabricated usually from the elements Silicon (Si) or Germanium (Ge). PV technology applications vary from powering a scientific calculator to powering a million houses per PV station, such as the Benban solar station that has been connected to the grid in late 2019 as the largest PV station in Africa, powering about 1,000,000 houses [4].

Solar PV technologies can be primarily divided into three types, such as:

1. Crystalline PV—monocrystalline and polycrystalline
2. Thin-film PV or amorphous PV
3. Emerging PV—organic PV and perovskite PV

The formation of solar PV modules will be discussed in the next chapter. An in-depth discussion on each type of module is excluded, as this book intends to provide a comprehensive illustration of PV systems and their applications to familiarize the readers with minimal engineering or technical knowledge with the field terminology up to being able to do system sizing of any solar application anywhere in the world.

1.3   Advantages, Challenges, and Potential Solutions

Similar to any other technology in the world, PV technology has some distinctive pros and cons. In the following list, some of the reasons to support the varying opinions on the technology with pertaining solutions to the challenges are given.

1.3.1 Advantages

There are numerous advantages of solar PV technology. These are delineated in the following paragraphs:

1. **Clean energy:** The electricity yield by the solar PV modules is not accompa-nied by any greenhouse gas (GHG) emission. GHGs are responsible for serious environmental damages such as global warming and climate change, and PV technology ensures energy generation is devoid of emission.
2. **Sustainable energy source:** The core of the PV technology is based on converting sunlight into electricity using modules made from Silicon, which is found in sand and among the most abundant elements on the Earth. As long as the sunlight hits the globe, we will have guaranteed availability of electricity.
3. **Noise-free technology:** PV systems usually do not contain any mechanical moving parts that may create noise. Therefore, it is considered a suitable renewable resource that can be installed in residential neighborhoods. However, other renewable generation systems are too bulky in size and complicated in operation to befit installation in localities.
4. **Low maintenance technology:** PV modules require very little technical main-tenance. Occasional cleaning is required to ensure that waste materials do not block the sunlight from reaching the modules.
5. **Performance warranty over operating lifespan:** PV module manufacturers usually provide a linear performance warranty on the modules in their datasheet.

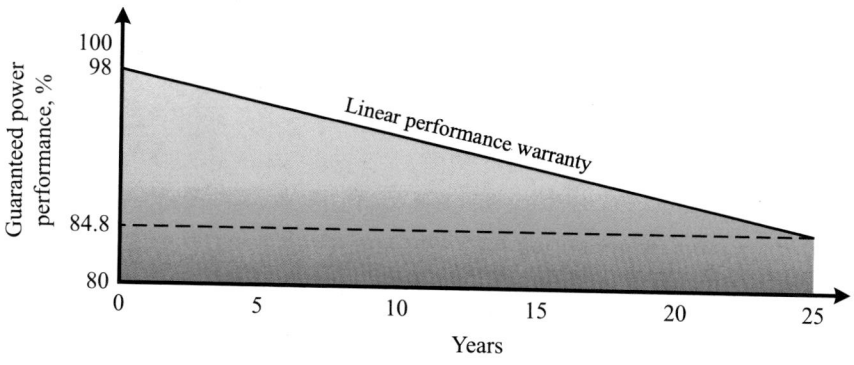

Fig. 1.2 The warranted power output of a typical solar module for 25 years

This warranty ensures the module's efficiency to be above a particular value during its operating lifespan. The warranty is advantageous as this indicates that the efficiency would drop very slowly over time. The graph in Fig. 1.2 demonstrates the warranted power output with respect to the years. As per the figure, the module operating efficiency will not decrease below 84.8% of its initial efficiency in its 25 years of operating lifespan. The initial efficiency is less than 100% due to the quality of the wafer manufacturing and crystal structure. After the module is exposed to the sunlight, about 0.5–1.5% module efficiency drops within the first few days depending on the types of modules. This unavoidable drop of efficiency is generally known as light-induced degradation (LID) [5].

6. **No geographical requirements:** PV modules can be installed in any part of the world regardless of any site condition since PV modules partially generate electricity even in cloudy locations. However, a sunny site free from shadows of tall trees or buildings is preferable for installing PV systems.
7. **Low cost:** During the last two decades, the cost of PV modules has dramatically declined to reach 0.214 $/W as an ex-factory gate price for global polycrystalline Silicon modules. Rapid advancement in recent semiconductor technologies indicates further price reduction in the near future.
8. **Mature but still promising:** Even though the PV industry has boomed at the beginning of the twenty-first century, the technology has shown relatively good maturity and potential in penetrating the well-established utility grids industry with a high development margin in the upcoming years regarding increasing efficiencies and lifetime.
9. **Recycling of solar modules:** Each solar module has a particular lifespan after which the module can be sent for recycling as it is mainly composed of glass, Silicon, and some metal.
10. **Less risk from natural disasters:** Solar PV systems are comparatively less vulnerable to natural disasters such as earthquakes, floods, and tsunamis [6].

Other power generation systems are more susceptible to damage from these disasters.

1.3.2 Challenges

There are some particular challenges associated with the installation of solar PV technology as well, which are mentioned hereunder:

1. **High initial and recycling cost:** PV applications have a relatively negligible maintenance cost. However, the initial cost is sometimes high in comparison to conventional resources for large-scale applications. Moreover, at the end of usage, executing several recycling processes such as separation, melting, and breaking each module might cost more than its selling price.
2. **Unreliable and fluctuating energy:** Energy yield from solar modules is proportional to the unstable solar irradiance that is available during the daytime, which also varies due to different shading. Therefore, solar energy is sometimes considered secondary in the electricity markets for its sporadic nature.
3. **The necessity for energy storage:** As a solution for the intermittency of solar power, energy storage technology introduced a way to compensate for the fluctuation during the shading periods. However, the lifespan of the energy storage elements such as batteries is shorter than the lifespan of the solar modules, and these elements must be changed multiple times throughout the project's lifespan. In other words, a new operating cost is sometimes included in the PV system, causing a rise in the initial cost to the PV system.
4. **Toxicity and disposal issue:** The process of manufacturing the PV modules is not fully carbon emission-free. The inclusion of some heavy elements such as Cadmium, Lead, and Selenium in some solar modules may spread toxicity when dismantled. Even though solar modules can be recycled, some parts cannot be completely recycled, producing additional wastage.
5. **Occupied area:** Since the efficiency of the light-electricity conversion process is relatively low, PV modules must cover a large surface area to meet the load or grid demand for energy. This challenge is related to the installation location and cost of the land.

1.3.3 Potential Solutions

Since solar technology is rapidly evolving due to its large usage in many parts of the world, researchers and engineers are developing solutions to overcome its challenges. With the commercialization of the latest semiconductor technologies, the cost of solar modules is decreasing significantly. Although the initial cost is high, the system offers the return of investment within a short period in most cases. In addition, energy storage systems such as batteries are being integrated to address the issue of unreliable and fluctuating energy. Optimized controllers are being employed with these batteries to increase their lifetime. Researchers are also working on making recycling and disposal of the expired module parts less expensive and avoiding the emission of toxic materials. Local and federal regulations on proper disposal of hazardous material are being enforced to avoid toxicity and disposal issues. Moreover, innovative mounting mechanisms provide the installers with options to make optimum use of the lands. Continuous research and advancement in solar technologies will substantially diminish the challenges faced in solar module installation, making it one of the leading energy sources in many parts of the world in the future.

1.4   PV-CSP Hybrid Technology

Thermal or heat energy is the major share of the obtained energy from sunlight. Solar thermal technologies, also known as concentrated solar power (CSP) technologies, aim to capture this huge amount of heat energy from sunlight and use it for heating applications or generating electricity. Due to the large size of the PV-CSP hybrid plants, they are typically used for large-scale energy generation [7]. Therefore, the detailed design process of large-scale solar energy technologies is not included in this book. But since hybrid solar plants are increasingly gaining momentum throughout the years, such plants are briefly described in the subsequent sections.

Two prominent CSP technologies used in conjunction with the PV technology are the parabolic trough and the solar power tower. A hybrid PV-CSP plant often incorporates energy storage systems, particularly thermal energy storage (TES) for the CSP facility and battery energy storage (BES) for the PV facility, for storing the energy for periods of solar intermittency. Figure 1.3 demonstrates the schematic diagram of the solar PV-CSP technology, where the PV plant and CSP plant coexist, and the battery bank and power block provide the required energy to the load.

1.4.1   PV-CSP Project Examples

The Stillwater triple hybrid power plant [8] in Nevada, USA, is the world's first power plant with three RE technologies working side-by-side to generate power and

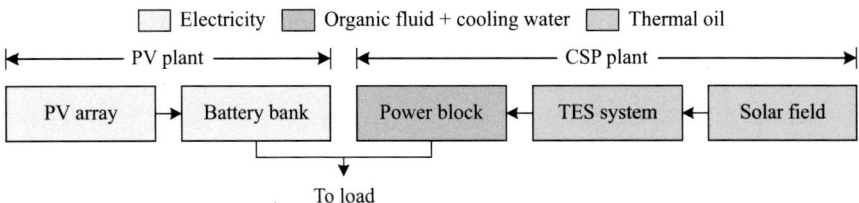

Fig. 1.3 Schematic of the solar PV-CSP hybrid technology

maximize the utilization of freely available natural resources. The plant contains a 33 MW geothermal plant, a 2 MW_e (17 MW_{th}) CSP plant as parabolic troughs, and a 26 MW solar PV plant with 89,000 solar PV modules. With a combined generation capacity of 61 MW, the project produces 15 GWh of energy each month.

The plant initially had only a binary cycle geothermal plant in 2001, based on the organic Rankine cycle. But a problem arose in the summer midday hours when the ambient temperatures went too high and reduced the output of the geothermal unit. As a result, the PV plant spanning 240 acres was added in 2012 to increase the geothermal output. Later, a CSP facility was also added in 2015 to increase the efficiency and output of the plant.

In South America, the Atacama Desert in northern Chile is ground to several PV-CSP hybrid projects due to the high levels of solar radiation. The Atacama 1 solar project [9] is under construction by the Spanish company Abengoa Solar, and it features a 110 MW solar power tower surrounded by heliostat mirrors with dual-axis tracking and a 100 MW PV plant with 392,000 modules having single-axis tracking. It also has molten salt energy storage facility that can store energy for 17.5 h. It is able to supply 410,000 houses with clean electricity 24 h a day. The Atacama 2 solar project is next in line by the same company and in the same desert, featuring the same design and capacity of CSP and PV plants [10]. However, the Atacama 2 project will have 15 h of molten salt TES capacity.

The Copiapó Solar project [11] had also been proposed in the Atacama Desert by SolarReserve, with two 120 MW (130 MW gross) solar power towers, a 150 MW PV plant, and 14 h of molten salt TES. With a 390 MW nameplate capacity and a planned annual output of 1700 GWh energy with 3360 MWh_e of storage capacity, the future of the proposed project is uncertain at present, as SolarReserve ceased operations in 2020 [12].

Another large PV-CSP hybrid complex in the Atacama Desert is the Cerro Dominador project [13], with 100 MW PV capacity, and 110 MW CSP capacity utilizing power tower technology surrounded by 10,600 heliostats. The plant also has a molten salt TES facility.

In South Africa, the Redstone project by ACWA Power [14] is planned to have a 271 MW PV-CSP plant with 100 MW of CSP capacity with 12 h of molten salt TES, and two PV plants rated 75 MW and 96 MW. With an initial investment of about 800 million USD, this project is the largest renewable energy investment in South Africa so far and promises to deliver continuous, clean electricity to 200,000 households. Construction of this plan began in May 2021 and is expected to begin commercial operations at the end of 2023 [15].

Notice that nearly all PV-CSP hybrid projects have their design in common solar power tower, PV, and molten salt TES capacity.

1.5   Historical Development of Solar PV

There were several attempts to take advantage of the Sun as a heat source, starting from focusing solar beams using a magnifier to set a fire until 1866 when the French scientist Augustin Mouchot started using the first parabolic solar collector to concentrate solar irradiance into boil water and generate steam that gives the kinetic energy to a steam engine.

A teenage French scientist Alexandre Edmond Becquerel discovered the photoelectric effect, during the industrial revelation in 1839. The first PV experiment was when he immersed a platinum chloride into an acidic conductive solution and noticed that when the setup is exposed to light, the result was a voltage boost. The setup was impractical but was efficient enough to open the door for other scientists to dig deeper into this field.

In 1876, Professor Willian Grylls and one of his students exposed sunlight into a Platinum and Selenium junction. Through the experiment, they proved that it is possible to obtain electricity from solid materials without chemical reactions. In New York City, the American inventor Charles Fritts improved the conversion efficiency to 1% by using a gold-Se junction that was the core of the first PV module installed on a rooftop in New York City in 1884.

The reason behind the photoelectric effect was not understood clearly until 1921, when Albert Einstein provided the theoretical explanation of it. Because of his outstanding discovery, he was awarded the prestigious Nobel Prize in the same year. In 1954, the game-changing step of the PV industry took place in Bell Laboratories in the United States, where researchers D. M. Chapin, C. S. Fuller, and G. L. Pearson realized that impurity-doped semiconductor materials could generate electricity when being exposed to light with a relatively high efficiency that reached 6%. Since then, semiconductor materials have replaced the Selenium-based solar modules.

The applications on the Earth were limited to powering remote telephone stations and small loads. However, due to the space race between the USA and the Union of Soviet Socialist Republics, PV technology applications were mainly used to power satellites orbiting the globe. In 1958, the USA launched Vanguard I, the first solar-powered satellite that weighed no more than 3.5 pounds.

The financial crisis caused by the oil crisis in the 1970s was the ignition of tremendous governmental support to RE, especially solar energy, to transfer it from being used in space applications to being a well-developed and mature technology in Earth-based applications. Therefore, in 1974, the US government announced establishing the National Renewable Energy Laboratory (NREL) as the first step toward this goal. Table 1.1 shows the evolution of PV in different ages.

Table 1.1 The chronological growth of the PV industry around the world [16–18]

Year	Event
Early period	
1839	Discovery of the PV effect by French scientist Alexandre Edmond Becquerel
1873	Discovery of PV effect in Selenium by Willoughby Smith
1876	Discovery of PV effect between the junction of Platinum and Selenium on illumination by William G. Adams with student Richard E. Day
1883	Production of first solar cells made of Selenium wafers by an American inventor Charles Fritts
1884	Installation of first solar cell made of Selenium wafers on New York rooftop by Charles Fritts
1887	Discovery of the fact that "ultraviolet light can change the voltage which would initiate sparks between two metal electrodes"
Medieval period	
1904	Theoretical explication of PV effect given by Albert Einstein in his scientific journal, *Annalen der Physik* (Annals of Physics)
1916	Corroboration of Einstein's theory by Robert Millikan through his measurements and obtaining an accurate value of Planck's constant
1918	Discovery of monocrystalline Silicon production by Jan Czochralski
1932	Observation of CdSe exhibiting PV effect
1941	Establishment of the first monocrystalline Silicon solar cell
1951	Establishment of first Germanium solar cells
1953	Derivation of solar cell efficiency by Dan Trivich
1955	Initiation of using solar cells to build satellite energy supply and demonstration of "the world's first solar-powered automobile" by William G. Cobb
1957	Establishment of 8% efficient solar cell by Hoffman Electronics
1958	Establishment of 9% efficient solar cell by Hoffman Electronics and launching of Vanguard-I, the first satellite to have solar electric power
1959	Establishment of commercially obtainable 10% efficient solar cell by Hoffman Electronics
1960	Establishment of 14% efficient solar cell by Hoffman Electronics
Modern period	
1962	Launching of Telstar, the first telecommunication satellite covered with solar cells with power generating capability of 14 W
1963	Establishment of first functioning PV module by using Silicon solar cells
1964	Launching of first Nimbus robotic satellite by NASA, capable of running solely on 470 W solar array
1968	Launching of OVI-13 satellite having two CdS solar modules
1969	Establishment of Spire Corporation, industry that produces equipment of solar cells
1973	Establishment of Solar One, a photovoltaic-thermal hybrid system in Delaware University
1975	Establishment of Solec International and Solar Technology International
1976–1985	Installation of PV systems on the Earth for many applications, such as power supply for water pumping, refrigeration, and lighting
1977	Production of PV modules rated more than 500 kW worldwide

(continued)

Table 1.1 (continued)

Year	Event
1979	Installation of the first PV on the White House by President Jimmy Carter, which was removed after a few years
1980	Establishment of first PV modules producing 1 MWp power per year
1981	Establishment of Solar Challenger, the first solar-powered aircraft by Paul MacCready
1982	PV modules surpassed 9.3 MW production worldwide
1983	PV modules surpassed 21.3 MW_p production worldwide
1984	Establishment of first amorphous PV modules
1985	Establishment of more than 20% efficient solar cells at the University of New South Wales in Australia
1986	Establishment of the first solar cell having thin film
1990	Establishment of the United Solar Systems Corporation for solar cell production
1992	Placement of a 0.5 kW PV system for a laboratory system in Antarctica
1994	Establishment of the first monolithic two-terminal solar cell of more than 30% efficiency by NREL using GaInP/GaAs
1999	Total cumulative worldwide volume of solar cells surpassed 1000 MW_p
2001	HELIOS, a solar-powered plane, broke the record by reaching a height of 30 km
2003	Connection of the world's largest PV power plant (of that time) with the power grid of Germany, "Solarpark Hemau" of 4 MW_p power
2004	Establishment of large solar systems of up to 5 MW in Germany and demonstration of 28% efficient Concentration PV (CPV) module [19]
2006	Demonstration of 33% efficient dual focus High Concentration PV (HCPV) [20]
2007	Installation of a 15 MW PPA (power purchase agreement), Nellis Solar Power Plant using modules of SunPower Corp
2011	Depletion of the manufacturing cost of PV modules to $1.25 per watt in China
2013	Installation of solar PV exceeds 100 GW worldwide and demonstration of 35.9% efficient HCPV
2014	Re-installment of PV solar modules in White House by the instruction of President Barack Obama
2016	Establishment of 34.5% efficient solar cell at University of New South Wales in Australia
2017	Installation of 300 GW PV solar modules globally at the beginning and exceeding 401 GW by the end of the year
2018	Installment of 480 GW solar PV globally
2019	Installment of 580 GW solar PV globally
2020	Installment of 707 GW solar PV globally

1.6 Growth of Solar PV

Starting from the first decade of the twenty-first century, most of the developed countries set RE transformation plans, also known as national RE portfolio goals. Since then, the PV industry has achieved a compound annual growth rate (CAGR)

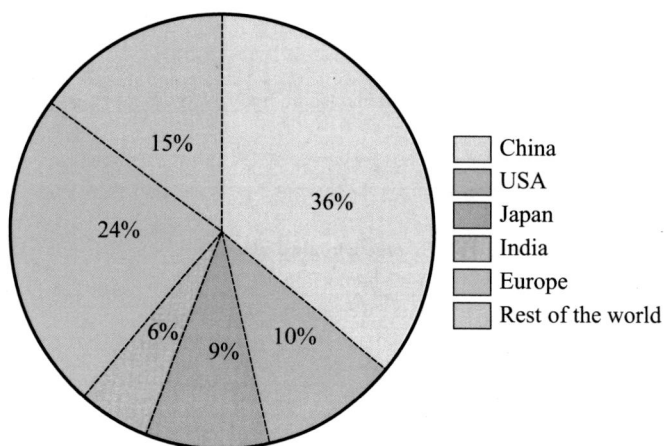

Fig. 1.4 Global cumulative PV installation by region in the year 2019 [21]

of 36.8% between 2010 and 2018. This high rate reflects the importance of the PV industry in the global energy vision, especially after the increasing warning alerts of the catastrophic environmental impacts of the greenhouse gases that are found from the emission of fossil fuel-based power plants. As a result, the rise of global PV integration has significantly increased. A summary of the rise is demonstrated in Fig. 1.4. According to the figure, China seems to be dominating the total global PV production to date [21].

The massive production percentage of solar PV modules reached 36% of the global cumulative PV installation and amounted to 480 GW_p by the end of 2018. This is the second highest installed capacity of RE source after wind energy that year. The trend continued as the number of global PV installations reached 580 GW_p by late 2019.

In Fig. 1.5, it is seen that Asia dominated the global PV installed capacity in 2018, followed by Europe and then North America, Oceania, the Middle East, Africa, and Latin America. However, based on the International Renewable Energy Agency (IRENA)'s transformation roadmap, North America is expected to exceed Europe by 2030 to be the continent with the second highest installed PV capacity. The installed capacity of the PV system worldwide is expected to increase to 2840 GW by the next decade and is anticipated to reach 8519 GW by 2050.

The new technological developments in the manufacturing process and in extracting the raw materials are necessary for solar PV modules. Moreover, the increasing investments in the industry can play a major role in decreasing the levelized cost of electricity (LCOE) produced by the solar PV modules to make solar PV the cheapest and most dominant RE source by 2050. This can be comprehended from Table 1.2, which shows the different factors that constitute the overall growth of the PV technology.

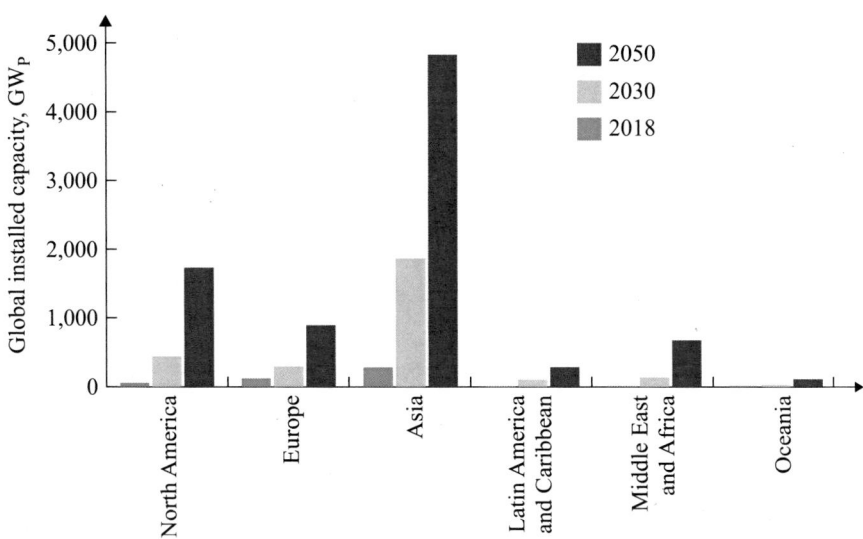

Fig. 1.5 Solar PV installed capacities (GW) around the world by 2018 (real-time data), 2030, and 2050 (estimated data) [22]

Table 1.2 The growth of solar PV technology from 2010 to 2018 (real data) and from 2030 to 2050 (predicted) [22]

Growth parameter	2010	2018	2030	2050
Share of solar PV in total generation, %	0.2	2	13	25
Total installed capacity, GW	39	480	2840	8519
Annual additions, GW/year	17	94	270	372
Total installation cost, $/kW	4621	1210	834–340	481–165
LCOE, $/kWh	0.37	0.085	0.08–0.02	0.05–0.01
Average annual investment, billion $/year	77	114	165	192
Jobs created, million	1.36	3.6	11.7	18.7

1.7 Cost of Solar PV

The cost of solar PV technology has always been showing a declining trend [3]. PV system cost terminology varies according to the application category and the components required to complete the system architecture. Thus, we will focus only on the PV module cost, which is the most critical portion of any PV system. Let us consider the cost drop of the global polycrystalline Silicon modules in the global market from 2014 to 2019. The cost of such modules started at 0.67 $/W in the first quarter of 2014, and it went down to 0.214 $/W in the second quarter of 2019—a 68% drop in only 5 years. The percentage of the global annual production of polycrystalline modules (60.8%) almost doubled the monocrystalline's (32.2%)

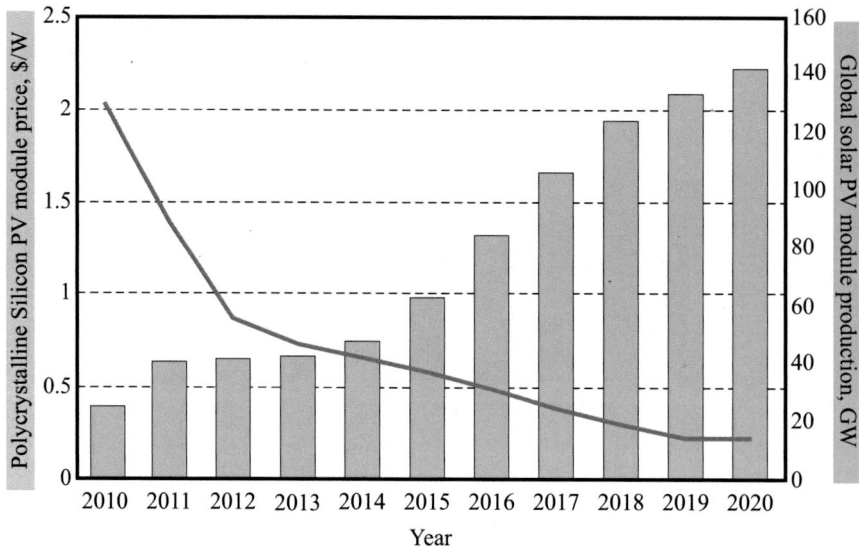

Fig. 1.6 The production of polycrystalline silicon solar PV cells has increased considerably in the last decade alongside the sharp decline in the module cost [23]

[21]. The production of polycrystalline silicon solar PV cells and the module cost in the last decade are illustrated in Fig. 1.6.

It is also possible to compare the financial feasibility of the unit of electricity generated by different energy sources using the techno-economic terminology LCOE. The LCOE calculates the average net expense of the unit of electricity (usually kWh) for any power generating plant over its expected lifetime. It is measured in $/kWh.

Studies show that in 2018, the electricity generated from solar PV power plants was already competitive to fossil fuel-based power plants. According to the International Renewable Energy Agency (IRENA), the LCOE range of the electricity generated by fossil fuel was 0.05–0.17 $/kWh, while the grid-connected solar PV LCOE varied from 0.06 to 0.21 $/kWh. In 2010, the LCOE of solar PV was 0.37 $/kWh, and that of onshore wind was 0.08 $/kWh. Within 10 years, the LCOE of solar PV came down to 0.085 $/kWh, and that of onshore wind was 0.06 $/kWh. Figure 1.7 shows the variation of the LCOE of solar PV and onshore wind power from 2010 to 2050. In most cases, the LCOE of solar PV is much higher than that of onshore wind. However, the LCOE seems to undergo many drastic reductions for solar PV compared to onshore wind, and if this trend continues, solar PV may even exceed wind energy. The predictions for 2030 show that the LCOE of solar PV will drop to reach the range of 0.02–0.08 $/kWh and that of onshore wind energy will reach the range of 0.03–0.05 $/kWh, and both of them are considered more economically feasible than fossil fuel, whose price will stay more or less constant. Further predictions to 2050 say that the LCOE of solar PV will reach

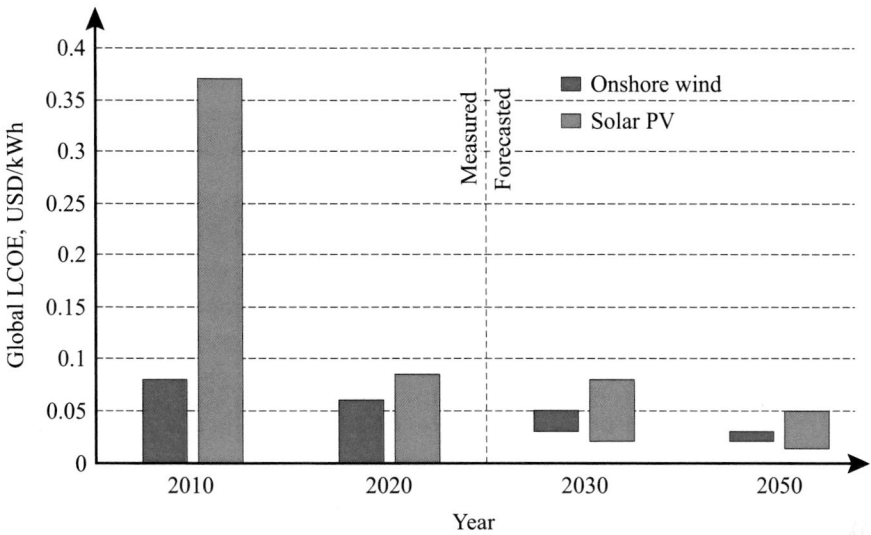

Fig. 1.7 The drop of global levelized cost of electricity (LCOE) in USD per kWh for solar PV and onshore wind power, connected with the utility grid [22, 24]

0.014–0.05 $/kWh, and that of onshore wind energy will reach the range of 0.02–0.03 $/kWh.

Once the initial system installation is complete and the capital costs are paid off, minimal maintenance charges are applicable in a solar PV system. Moreover, with government incentives, subsidies, net metering systems, time of use billing systems, and other advantages, the overall lifetime costs are much lower for solar PV systems compared to conventional power generation systems.

To summarize, the cost of solar PV systems significantly declined over the year, having lower costs compared to the existing fossil fuels. Similarly, the LCOE of the solar modules has also decreased considerably with respect to coal and natural gas and even compared to other renewable options such as wind and geothermal energy.

1.8   The Efficiency of Solar PV

The efficiency of any energy conversion apparatus is the ratio of output power to input power. In solar PV modules, the ratio of the output electrical power provided by the module to the solar energy reaching the PV module's surface is the efficiency, as represented in Eq. 1.1.

$$Efficiency, \eta = \frac{Output\ electrical\ power}{Input\ solar\ power} \times 100\%. \tag{1.1}$$

At present, the efficiency of single-junction solar PV cells ranges from 15 to 28%, with a maximum theoretical efficiency of 33.16% [25]. This implies that out of 100 Watts of energy incident on the solar module from the Sun, for example, only 15 to 28 Watts of energy are effectively converted into usable electrical energy, while the rest of the energy is lost as heat energy or lost through reflection and refraction.

Each solar module contains many PV cells that are connected in series or parallel to make a PV module. The theoretical efficiency of both the cells and modules can be calculated using the formula above. It is to be noted that the efficiency of the module is less than that of the efficiency of the cell. It is because as the cells are connected to each other, the spacing increases due to additional mounting and accessories, which causes energy losses and reduces the module efficiency. The differences in the cell and module efficiencies are depicted in Fig. 1.8.

In 2020, the module efficiency record of the dominant PV technology (Silicon-based polycrystalline) belonged to Hanwha Q Cells. This South Korean PV module manufacturer succeeded in converting 20.4% of the sunlight's power into electricity, where a majority of the incident irradiance was lost and not converted into valuable energy.

The efficiency of solar PV systems is on the rise over the years, thanks to tireless research and development underway. Figure 1.9 demonstrates the rise in efficiencies of different types of PV systems, i.e., concentrated solar PV systems or concentrators, crystalline Silicon, and thin-film PV systems over the years. It is evident from the figure that the efficiency of the concentrators has increased significantly, from about 14% in 1995 to 25% in 2020. The efficiency of these modules is prognosticated to increase in the coming years.

The increase in efficiency will also correspond with the decreasing prices of PV systems in the near future. However, the efficiency of practical solar PV systems

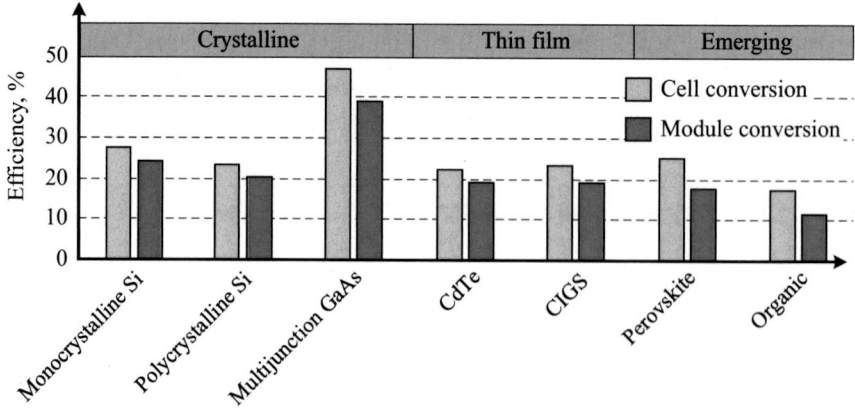

Fig. 1.8 The efficiencies of the different types of PV cells and modules. It is seen that the efficiency of the cells is greater than the corresponding modules [26]

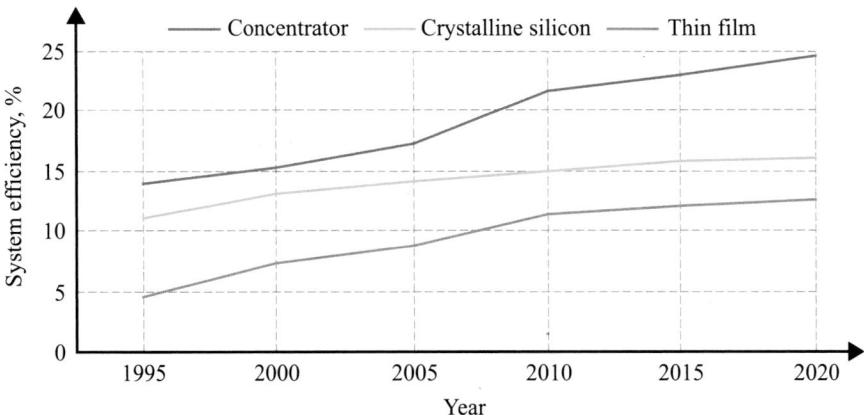

Fig. 1.9 The efficiency of solar PV systems has been increasing with a positive slope [27]

varies from the test conditions due to several factors such as orientation, tilt, shading, aging, overheating, and so on. These factors will be dealt with in later chapters.

1.9 Required Skills to Install Solar PV Systems

The range of the PV industry is very wide; it can be as small as installing a few solar modules to power numerous small home appliances up to the design, installation, and integration of several hundreds of MW of solar PV modules to form a utility-scale PV plant that is usually connected to a medium or high voltage interconnection of the existing electricity grid through several inverters and transformers. Therefore, to be competitive in this field, several skills are needed to achieve the optimum design of any size with the least possible expenses.

Since PV modules are electricity-generating devices, electrical power-related expertise is of considerable importance in this context. However, the complete integration of the PV systems incorporates mechanical, civil, structural, electronics, and economic aspects as well. The insights of these departments are equally important to play a role in the design of a full-fledged solar PV system.

1.9.1 Electrical Engineering Skills

Since solar PV systems are designed for the generation of clean electricity, most of the work and responsibilities are for the electrical department of the company. Some of the notable responsibilities can be listed as follows:

- Select the appropriate system components such as PV modules, inverters, transformers, batteries, and cables.

- Build the system based on the most economical components that meet the national code standards.
- Optimize the PV system sizing based on the customer needs or the utility requirements, patterns, and rates.
- Support the electrical and mechanical structure with a protection system that includes lightning and grounding protection, circuit breakers, fuses, and switches to assure the safety of the people and the equipment to guarantee the easiest maintenance operations in the future.
- Provide the best understanding of the interconnection conditions of each utility grid.
- Draft software-based single-line, three-line diagrams with the module design and roofing plans for the proposed electrical wiring.
- Perform all necessary calculations to avoid shading on the modules; if not possible, cooperate with the mechanical, civil, and structural engineers and the economic specialists to determine the best row spacing of the system.
- Develop professional skills in residential, commercial, and industrial PV software tools to provide accurate work.
- Provide all required documents and calculations to the customers or utility grid representatives.
- Estimate monthly and annual parameters such as energy output, return of investment, and amount of carbon footprint reduction for each project to give the investors or the customers a clear idea about the technical and financial feasibility of the project.

1.9.2 Mechanical, Civil, and Structural Engineering Skills

The installation of solar PV systems requires mechanical, civil, and structural engineering skills as well. To name a few, the tracking mechanism, the orientation, and the cooling of the modules are under the scope of mechanical engineering, whereas the installation and site allocation are under the scope of civil and structural engineering. The following are some specific responsibilities of these departments in the domain:

- Receive the estimated number of modules from the electrical engineers and select the appropriate mounting systems and constructible areas based on the geo-technical reports.
- Analyze the data received from the electrical engineers to optimize the construction aspects, including mounting structure and draining systems of the rain or stormwater.
- Acquire and analyze wind data of the previous years to determine the maximum wind load that the mounting structure must stand.
- Estimate and consider snow loads based on historical data.
- Provide all necessary technical designs or software diagrams and documentation to the customer, investor, or utility representative.

1.9.3 Electronic and Programming Skills

The fabrication of solar PV modules and the development of the PV technology are largely dominated by electronic engineers, and the automation of the tracking and cooling systems requires good programming skills for reliable operation. The following are some noteworthy electronic and programming skills required for solar PV systems:

• Design, program, and calibrate the tracking system that consists of motors and their drivers run by an algorithm based on time or photocells.
• Monitor the output and apply communication protocol between the monitoring stations and the power plant using online tools and mobile applications.
• Select the appropriate setting of all electronic components that guarantees the best use of the whole system.

1.9.4 Economic Skills

Optimizing the cost and maximizing the profit through rigorous economic analysis while designing the solar PV system are the important aspects. Hence, the solar PV design engineer must have economic skills for optimizing the design. The following are some of the economic responsibilities of the designer:

• Gather the necessary data about the land cost, annual expected electricity generation, maintenance schedule and expenses, and grid rates to perform the techno-economic analysis of the system and determine the economic feasibility of the plant.
• Compare PV with alternative solutions such as diesel generators for off-grid applications to convince the customers of the PV option.

1.10 Conclusion

This chapter introduces solar energy and PV technology. It also elaborates how solar PV meets electricity requirements by promoting an eco-friendly environment. The continuous rise of electricity demand attracts researchers to tackle the emerging challenges. This widely distributed PV system has been described in terms of costs and global installations in this chapter. The depletion of cost over the development of PV has been discussed, comparing historical studies with present and future analysis through graphical representation. According to the development analysis, the prediction of future development has been manifested in terms of cost, capacity, and electricity production. Finally, the skills required to design a PV have been discussed by dividing them with respect to different fields. This chapter will assist the readers in gaining a basic understanding of PVs, which will aid the understanding of a more in-depth study on the system in the subsequent chapters.

1.11 Exercise

1. What are the different ways to harvest solar energy?
2. What is meant by solar photovoltaic technology?
3. What are the advantages, challenges, and associated solutions of solar photovoltaic technology?
4. What is the input power of a 20% efficient solar PV module whose output power is 300 W? [Answer: 1.5 kW]
5. Why is the cost of solar PV systems decreasing over the years although the efficiency is increasing?
6. Briefly elaborate the different types of skills required to design and install a PV system.

References

1. Solar Energy Industries Association, Solar Technologies (2021). https://www.seia.org/initiatives/solar-technologies
2. Solar Energy Water Heating System Monitoring (2013). https://www2.advantech.com/ia/newsletter/AutomationLink/July2013/RU/appstory02.html
3. E. Hossain, S. Petrovic, Renewable Energy Crash Course (2021)
4. J. Wells, Benban, Africa's largest solar park, completed (2019). https://www.ebrd.com/news/video/benban-africas-largest-solar-park-completed.html
5. A. Gong, Understanding PV system losses, part 1: nameplate, mismatch, and lid losses (2021) https://www.aurorasolar.com/blog/understanding-pv-system-losses-part-1/
6. E. Hossain, S. Roy, N. Mohammad, N. Nawar, D.R. Dipta, Metrics and enhancement strategies for grid resilience and reliability during natural disasters. Appl. Energy **290**, 116709 (2021)
7. X. Ju, C. Xu, X. Han, H. Zhang, G. Wei, L. Chen, Recent advances in the PV-CSP hybrid solar power technology, in *AIP Conference Proceedings*, vol. 1850 (AIP Publishing LLC, 2017), p. 110006
8. G. DiMarzio, L. Angelini, W. Price, C. Chin, S. Harris, The Stillwater triple hybrid power plant: integrating geothermal, solar photovoltaic and solar thermal power generation, in *Proceedings World Geothermal Congress* (Citeseer, 2015), pp. 1–5
9. Abengoa Solar, Solar complex atacama 1 (2021). http://www.abengoa.cl/web/en/areas-de-actividad/energia-solar/proyecto/Solar-complex-Atacama-1/
10. Abengoa Solar, Abengoa Solar (2021). http://www.abengoa.cl/web/en/areas-de-actividad/energia-solar/abengoa-solar/
11. P. Fairley, Chile's hybrid PV-solar thermal power stations (2015). https://spectrum.ieee.org/energywise/green-tech/solar/chiles-hybrid-pvsolar-thermal-power-stations
12. Reve, The closure of SolarReserve, an isolated case of the concentrated solar power industry (2020). https://www.evwind.es/2020/02/07/the-closure-of-solarreserve-an-isolated-case-of-the-concentrated-solar-power-industry/73461
13. Renewables Now, Chile's Cerro Dominador CSP plant synced with the grid (2021). https://renewablesnow.com/news/chiles-cerro-dominador-csp-plant-synced-with-the-grid-737626/
14. ACWA POWER, Redstone CSP IPP (2021). https://acwapower.com/en/projects/redstone-csp-ipp/

15. Solar Paces, 100 MW Redstone solar with 12 hours daily thermal energy storage closes financ-
    ing (2021). https://www.solarpaces.org/100-mw-redstone-solar-with-12-hours-daily-thermal-
    energy-storage-closes-financing/
16. Denis Lenardic, Photovoltaics—Historical Development (2015). https://www.pvresources.
    com/en/introduction/history.php
17. L. Fraas, *Low Cost Solar Electric Power*, chapter 1 (Springer, Berlin, 2014). https://www.
    researchgate.net/publication/274961489_Chapter_1_History_of_Solar_Cell_Development
18. Renewable Capacity Statistics 2021 (2021). https://www.irena.org/publications/2021/March/
    Renewable-Capacity-Statistics-2021
19. K. Araki, M. Kondo, H. Uozumi, Y. Kemmoku, T. Egami, M. Hiramatus, Y. Miyazaki, N.J.
    Ekins-Daukes, M. Yamaguchi, G. Siefer, et al., A 28% efficient, 400x and 200 wp concentrator
    module, in *19th European PVSEC* (2004), pp. 2495–2498
20. L. Fraas, J. Avery, H. Huang, L. Minkin, E. Shifman, Demonstration of a 33% efficient
    Cassegrainian solar module, in *2006 IEEE 4th World Conference on Photovoltaic Energy
    Conference*, volume 1 (IEEE, Piscataway, 2006), pp. 679–682
21. Fraunhofer Institute for Solar Energy Systems, Photovoltaics Report (2020). https://www.ise.
    fraunhofer.de/content/dam/ise/de/documents/publications/studies/Photovoltaics-Report.pdf
22. International Renewable Energy Agency (IRENA), Future of Solar Photovoltaic: Deployment,
    Investment, Technology, Grid Integration and Socio-Economic Aspects (A Global Energy
    Transformation: Paper) (2019). https://irena.org/-/media/Files/IRENA/Agency/Publication/
    2019/Nov/IRENA_Future_of_Solar_PV_2019.pdf
23. XIAOJING SUN, Solar technology got cheaper and better in the 2010s. Now
    what? (2021). https://www.greentechmedia.com/articles/read/solar-pv-has-become-cheaper-
    and-better-in-the-2010s-now-what
24. International Renewable Energy Agency (IRENA), Future of Wind: Deployment, Investment,
    Technology, Grid Integration and Socio-Economic Aspects (A Global Energy Transfor-
    mation: Paper) (2019). https://www.irena.org/-/media/files/irena/agency/publication/2019/oct/
    irena_future_of_wind_2019.pdf
25. NREL, Best Research Cell Efficiencies (2019). https://www.nrel.gov/pv/assets/pdfs/best-
    research-cell-efficiencies-rev210726.pdf
26. Center for Sustainable Systems, University of Michigan, Photovoltaic Energy Factsheet
    (2020). https://css.umich.edu/sites/default/files/Photovoltaic%20Energy_CSS07-08_e2020.
    pdf. Pub. No. CSS07-08
27. A. Han, Efficiency of solar PV, then, now and future (2015). https://sites.lafayette.edu/egrs352-
    sp14-pv/technology/history-of-pv-technology/

Fabrication of Solar Cell

2

2.1 Overview of Semiconductor Processing

For understanding the principle of the photovoltaic (PV) effect, it is essential to understand the physics of semiconductor processing first. The semiconductor is considered the core of PV technology. Using its inherent feature, the solar module can convert the exposed sunlight into a direct electrical current.

Conductivity is defined as the characteristic of a material to allow the flow of electric current through it. It is the inverse of the electrical resistance and measured in Ω^{-1} or Siemens. Based on electrical conductivity, materials can be divided into three main types—conductors, insulators, and semiconductors.

2.1.1 Conductors

Since the electrical current is a flow of electrons through a material, conductive materials show a minimum resistance toward the movement of electrons due to the redundant available free electrons in them. In other words, these materials facilitate the electronic movement from one atom to another once voltage is applied. Conductor materials, such as Aluminum and Copper, are mainly used to manufacture cables or motor windings. It is worth noting that all metals are considered good conductors of electricity and heat.

2.1.2 Insulators

Unlike conductors, insulators have shallow conductive properties, which means that they display a high resistive behavior toward electron flow through the material due to the low number of free electrons. Glass, rubber, ceramic, plastic, and some chemical alloys are a few examples of insulators. These materials are used as the

© The Author(s), under exclusive license to Springer Nature Switzerland AG 2022
Y. Abou Jieb, E. Hossain, *Photovoltaic Systems*,
https://doi.org/10.1007/978-3-030-89780-2_2

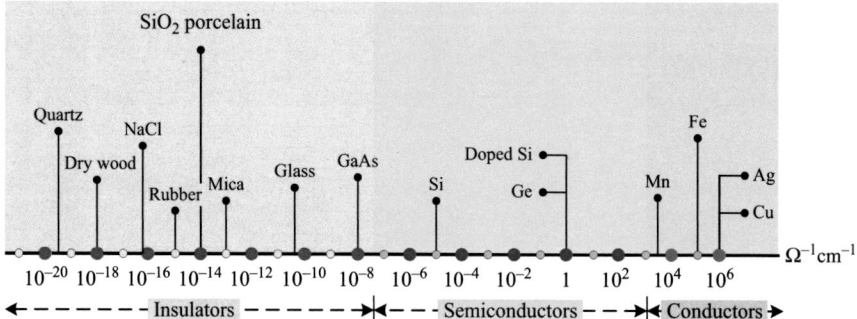

Fig. 2.1 Conductivity values of different types of materials

cover of the conductive materials in cable or motor applications to protect the users or maintenance personnel from any electrical shock. Insulators are also used in high voltage applications to separate the conductive lines and the metallic structure of the power transmission pole.

2.1.3 Semiconductors

Semiconductors such as Silicon (Si) and Germanium (Ge) can show both conductive and insulating behavior toward the flow of electrons based on the surrounding or internal circumstances. Temperature rise shifts the semiconductor material behavior toward conductive materials. Moreover, the intentional addition of impurity in materials to improve chemical performance through a process called doping is used to increase the material conductivity regardless of the temperature. The doping process is used in all semiconductor applications, such as in the manufacturing process of electronics (diodes, transistors, thyristors, etc.) and most importantly, in the context of this book, solar modules. Figure 2.1 illustrates the conductivity of a variety of materials used in the electricity field. Resistivity value of the materials can also be derived from the figure by obtaining the reciprocal of the conductivity [1].

Silicon is an essential material in the PV industry. This chapter will focus on Silicon process engineering, starting from Silicon as an atom until it is converted into a solar module.

2.2   Doped Silicon

Silicon is a chemical element with 14 electrons in its atom that fills the first and second shell with two and eight electrons, respectively. The last four electrons, known as valance electrons, remain in the outer shell. These four electrons allow the Silicon atom to bond with the other four Silicon atoms. This process prevents Silicon from being conductive because all valance electrons are bonded with electrons of other atoms. In other words, there are no free electrons to make the material conductive. Therefore, the doping process is introduced to provide the material with an extra electron or hole by adding some dopant materials such as phosphorous, arsenic, and gallium. Through doping, a controlled amount of impurities (dopants) is added on purpose within a pure semiconductor (also called intrinsic semiconductor) for altering its electrical, structural, and optical properties. Si doping is done by the intentional addition of trivalent or pentavalent atoms, involving chemical reactions so that the introduced impurities form a covalent bond by sharing electrons with the Silicon atoms.

2.3 Types of Doped Silicon

Silicon doping can be done by adding two types of impurities: trivalent atoms or pentavalent atoms. Therefore, doped Silicon can be classified into two types—n-type and p-type, based on the type of doping.

2.3.1 n-Type Semiconductor

n-Type Silicon results from replacing the Silicon atom in the Silicon lattice with a pentavalent atom, i.e., an atom of any chemical element containing 5 valance electrons. This process guarantees that an extra electron will be available in the material structure because four electrons of Silicon will bond with four valance electrons of the dopant atoms, leaving the fifth electron free as it has no electron to bond with. This translates to better electrical conductivity, as the free electron requires less energy to transition from the lower energy band, called valence band (VB), to the higher energy band, called conduction band (CB), allowing electric current to flow.

The dopant material, which is usually Phosphorus or Arsenic, is added in small amounts to the Silicon, which then donates an electron to the doped semiconductor. Thus, the doped semiconductor is then termed an n-type semiconductor. The negatively charged electrons become the majority carrier, and the relatively positive careers or holes act as the minority carrier, since the concentration of negative carriers is greater than the concentration of the positive carriers. The dopants used for n-doping are known as donor impurities as they donate electrons. Figure 2.2 illustrates the n-type doping of Silicon with Phosphorus, where the Phosphorus atom donates its fifth electron to be free as it has no electron to bond with. Thus, it acts as a free charge carrier.

2.3.2 p-Type Semiconductor

The p-type doping process is identical to the n-type doping in enhancing the electrical conductivity of the semiconductor but with a different dopant effect. In

Fig. 2.2 n-Type doping of
Silicon with Phosphorus

Fig. 2.3 p-type doping of
Silicon with Boron

p-type doping, the dopant will be a trivalent atom, i.e., an atom of a chemical element with three valance electrons such as Gallium or Boron. The three valance electrons of the dopant will bond with three corresponding electrons of Silicon atoms. As a result, the dopant can take an additional electron in its outer shell, which will create a hole in the VB of Silicon atoms. The electrons in the VB of Silicon atoms are mobile at this stage. It is to be noted that holes are positively charged and moves in opposition to the electron flow direction, raising the electrical conductivity of the doped semiconductor, also called p-type semiconductor. Holes do not move themselves; electrons move from one hole to another, which eventually translates to the holes moving in the direction opposite to that of the electrons. In p-type semiconductors, the majority charge carriers are holes, and the minority charge carriers are electrons. The dopants used for p-doping are known as acceptor impurities as they create holes and can accept an electron. A significant part of the current is produced due to the movement of holes in these semiconductors. Figure 2.3 illustrates the p-type doping of Silicon with Boron where Boron atom has a hole in its outer shell and accepts an electron from a Silicon atom. Note that the main purpose of the doping process is changing the hole–electron balance so that the Silicon is made more conductive [2].

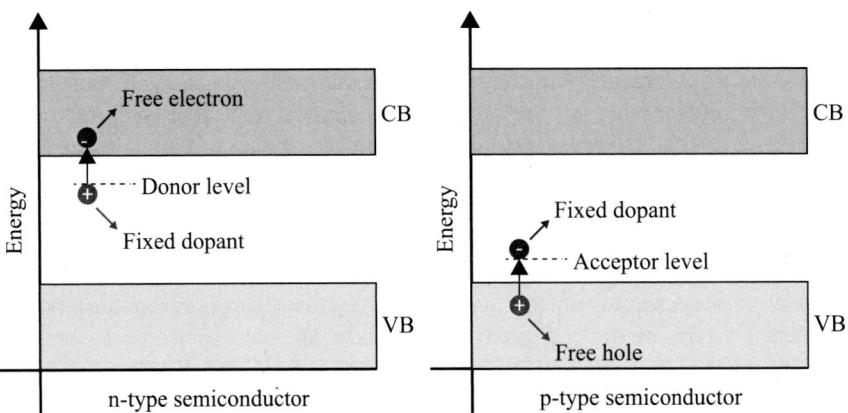

Fig. 2.4   Energy band structure of n- and p-type semiconductors

2.4   Energy Band Structure for n- and p-Type Semiconductors

The addition of impurities creates a localized energy level in between the VB and CB. The energy difference between the VB and the CB is the required energy to move electrons to the CB from the VB, and is typically known as the *bandgap energy*. For n-type semiconductors, the localized energy level is known as *donor energy level*. As discussed above, n-type semiconductors have a free electron that moves around, bonding with no other electron. Therefore, a small amount of energy is required to move the free electron to the CB, and for this reason, the donor level is closer to the CB, indicating a smaller bandgap.

In p-type semiconductors, the localized energy level of the holes of acceptor impurities is known as the *acceptor energy level*. By applying a small amount of energy, some electrons from the VB can be moved to the acceptor energy level, producing holes in the VB. If a high amount of energy is applied, more electrons will move to the CB, creating more holes in the VB. The acceptor energy level is situated slightly above the VB. Figure 2.4 illustrates the energy band structure of n- and p-type semiconductors [3].

2.5   The p-n Junction

A p-n junction is a sandwich of a p-type and an n-type semiconductor. The p-type and the n-type semiconductor can be formed by doping pentavalent and trivalent atoms into a layer of pure Silicon, respectively. The p-n junction is considered a very useful formation that allows us to take advantage of the unique character of the semiconductors, which is utilized in both the solar and electronics industries.

The p-type Silicon contains an excess of positive charge carriers, or holes, and some immobile negatively charged Boron atoms, also known as negative ions. The number of mobile charge carriers, or electrons, is exactly equal to the immobile Boron ions. Likewise, the n-type material contains many freely moving negatively charged electrons and the same number of immobile positively charged Phosphorus ions. Once the two materials are sandwiched together, some free electrons in the n-type layer migrate or diffuse toward the holes p-type material. Similarly, the holes diffuse toward the n-type layer.

The natural movement of the electrons and holes leaves the immobile positive Phosphorous ions in n-type (making the n-type layer slightly positively charged) and negative Boron ions in p-type materials (making the p-type layer slightly negatively charged). The two slightly positive and slightly negative charges across the junction (p-type and n-type interface) create an electric field. The holes in the p-type material repel the positive Phosphorous ions of the n-type material, and the electrons in the n-type material repel the negative Boron ions of the p-type material. This repulsion resulting in an evacuation or depletion of the interface area of both types from the free charges make a perfect insulator area called the *depletion zone or depletion region*. Moreover, a high concentration of holes in the periphery of the p-type and a high concentration of electrons in the periphery of the n-type is formed. The mentioned depletion region contains no free charges, which means it acts as a pure insulator exactly as any pure Silicon lattice.

The resistivity of the p-n junction might be easily controlled by connecting a voltage source in a forward voltage with any higher value than the electric field to get over the electric field of the depletion region and make the p-n junction a conductive material. This is known as a *forward bias* condition. In contrast, applying a reverse voltage by an external voltage source on the p-n junction will strengthen the electric field, which leads to higher resistivity for current flow in the p-n junction. This is known as a *reverse bias* condition. It is important to know that this technology is the reason behind the existence of electronic switches such as diodes, transistors, and thyristors that play huge roles in the engineering field.

All solar cells are made up of a p-n junction layer. Light is made up of tiny particles called photons, as proposed by Albert Einstein in 1905. The photons in the sunshine have enough energy to break some electron–hole pairs in the p-n junction, which creates free electrons and free holes. Once the electron receives enough energy from the photon, it gets excited and jumps into a higher energy state, leaving a hole behind. However, the excitation does not last very long, and a recombination process occurs with the surrounding hole while the energy of the electron dissipated as heat energy.

Once enough light is exposed to the solar cell, and electrons get excited, they can transition from the p-type layer, cross the depletion region, and go to the n-type layer, due to attraction to the slightly positively charge as mentioned before. If there is an externally connected electric load, the electron can flow from the n-type layer to the load through the external circuit and return to the p-type layer. It is essential to remember that the direction of electricity is reverse to the electron flow. Thus, as electrons move from n-type to p-type, the electricity flows from the p-type layer to the n-type layer [4]. The formation of the p-n junction is illustrated in Fig. 2.5, and the operation of the PV cell is depicted in Fig. 2.6.

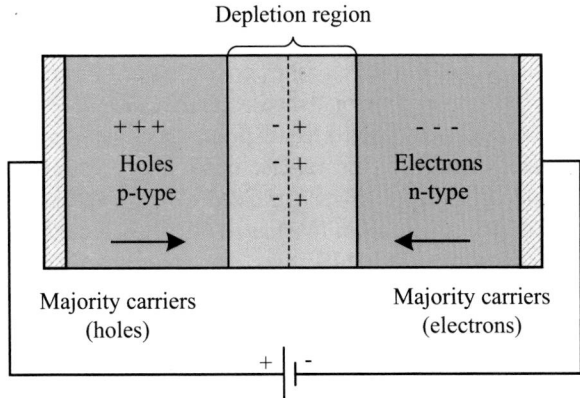

Fig. 2.5 Formation of the p-n junction by diffusing n-type and p-type Silicon

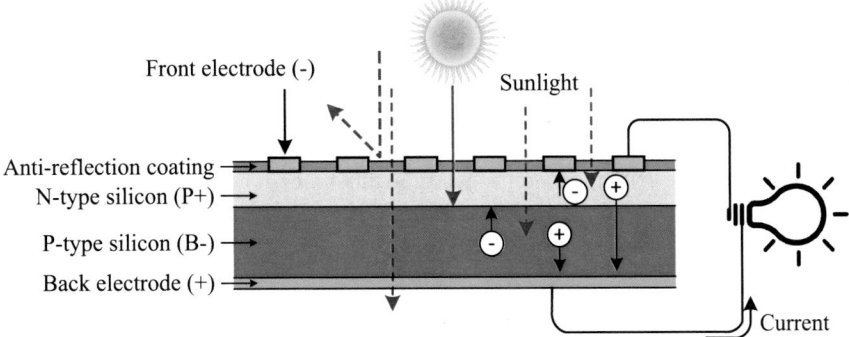

Fig. 2.6 Working principle of a PV cell connected to a bulb as an external load

A single cell can utilize only a particular fraction of the solar spectrum. Multiple cells with many p-n junctions, known as *multijunction (MJ) cells*, are capable of enhancing this limitation and extracting maximum energy from solar irradiation. These solar cells are also known as tandem cells since multiple p-n junctions are connected in tandem. MJ solar cells are also prevalent at present as they can significantly improve the efficiency of PV modules. Single-junction PV cells are only 33.5% efficient and are limited by the Shockley–Queisser (SQ) limit, i.e., the maximum hypothetical efficiency considering one p-n junction only. But MJ PV cells have demonstrated efficiencies over 45% under extreme conditions. If an infinite number of p-n junctions are used in MJ PV cells, the cell can theoretically reach an efficiency of 68.7% under normal sunlight and 86.8% under concentrated sunlight.

There are two types of MJ solar cells—monolithically integrated MJ and mechanically stacked MJ. The barrier between the p-n junctions, also called tunnel junctions, acts as ohmic electrical contacts and connects successive junctions in an

MJ solar cell. To elaborate, the tunnel junction is a barrier between the layers, and the electrons have to cross this barrier through a process called quantum tunneling. Elements from Group III and Group V of the periodic table are perfectly compatible for making compound semiconductors, such as GaAs, GaP, GaN, InAs, InP, etc. These materials have superior optoelectronic properties, including direct bandgap between the conduction band and the valence band, high electron mobility in the device, and low binding energy between localized electron and hole (exciton) [5]. Therefore, these materials are ideal for making III-V MJ solar cells. Due to the high efficiency and costs of fabricating MJ PV cells, they are primarily used in aerospace applications, such as in space shuttles and space stations.

2.6   Growing Ingots

A PV module is an array of many PV cells, and a PV cell is a simple p-n junction made of Silicon. In the upcoming sections, the chemical and physical process of manufacturing solar modules, from raw material to its final shape as a solar module that can be used by the end-user, will be discussed. The most common element on the Earth surface is Oxygen, followed by Silicon. Oxygen accounts for 46.6% and Silicon for 27.7% of the elements, followed by Aluminum, which makes up 8.1% of the elements. Silicon is available in the form of SiO_2 or Silica, which cannot be used directly in the semiconductor industry without a complex purification process that eliminates most of the impurities. Almost 99% pure Silicon is obtained from this purification process.

Low-quality Silicon called metallurgical grade polysilicon is obtained by a reduction process of SiO_2 available in the Quartz rocks. These rocks are melted in an arc furnace with a high temperature of about 1900 °C, and then carbon is added to the melted Silicon (Quartz) as shown in Reaction 2.1.

$$SiO_2 \rightarrow Si + 2CO. \tag{2.1}$$

Carbon monoxide is released in a gaseous state from the chemical process with almost 2% impure Silicon. This Silicon is found in melted condition at first, which later gets solidified. A further purification process known as the Siemens process (Fig. 2.7) is followed to convert the derived Silicon into trichlorosilane. This process is done by exposing the metallurgical grade polysilicon to hydrogen chloride (HCl) in a reactor with a high temperature of 300 °C, as illustrated in Reaction 2.2 .

$$Si + 3HCl \rightarrow SiHCl_3 + H_2. \tag{2.2}$$

The melted trichlorosilane is converted into polysilicon by passing the material through a reactor with hydrogen at a high temperature between 1000–1200 °C. The melted trichlorosilane will decompose, and HCl will be formed in addition to pure

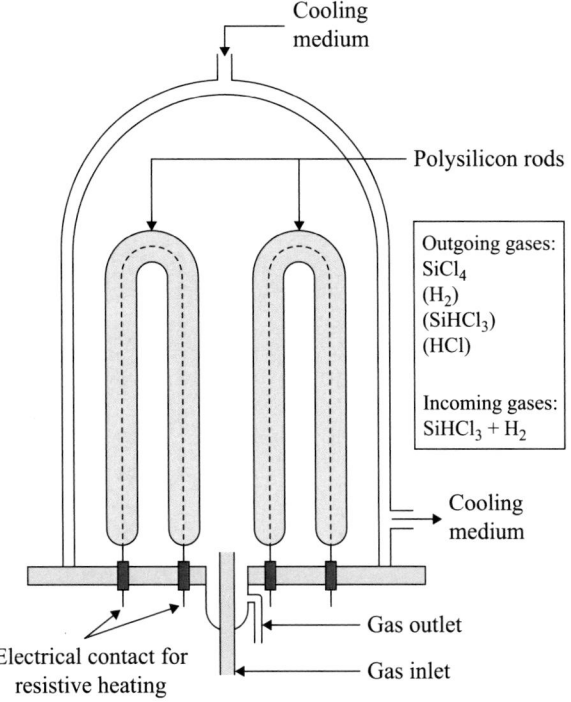

Fig. 2.7 Schematic diagram of a Siemens Reactor [4]

Silicon, called electronic grade polysilicon which can be found decomposed on the reactor rods. This is shown in Reaction 2.3.

$$SiHCl_3 + H_2 \rightarrow Si + 3HCl. \qquad (2.3)$$

It is worth noting that the Siemens process consumes a lot of energy due to using high temperatures. To overcome this high consumption of energy, another method is introduced to obtain electronic grade polysilicon with much less temperature, which means less energy consumption. Dynamic Silicon seeds are used in the system with silane and heated hydrogen streams. Small polysilicon granules are the byproduct of the system, and the unreacted silane and hydrogen are exhausted from the upper part of the system. This process is done using a Fluidized Bed Reactor (Fig. 2.8) which follows the reaction 2.4 as shown below:

$$SiH_4 + H_2 \rightarrow Si + 3H_2. \qquad (2.4)$$

The polysilicon obtained from both processes is considered to be highly pure, with an impurity ratio of less than 1 parts per million (ppm) Silicon particles. This

Fig. 2.8 Schematic diagram
of a Fluidized Bed Reactor
[4]

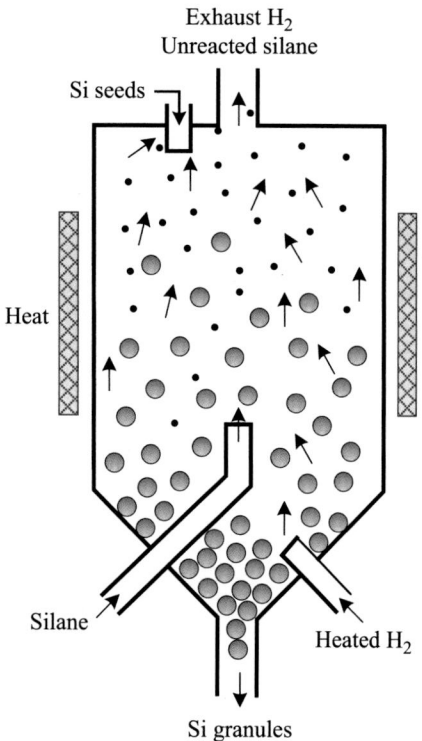

Silicon can be further melted and put in a temperature-resistive parallel rectangular
container to form a polycrystalline Silicon ingot.

The lattice of polycrystalline Silicon obtained from the above-mentioned processes contains discontinuous and disoriented crystals forming many grains. Solar
cells that are built based on this Silicon are called multi-crystalline or polycrystalline
solar cells, which can easily be recognized by the naked eye as a non-uniform dark
blue solar cell with boundaries [4].

In short, monocrystalline Silicon has continuous, uniform crystals having no
grains, whereas polycrystalline Silicon has discontinuous, non-uniform orientated
crystals with many grains. The grain boundaries are also visible in polycrystalline
Silicon (Fig. 2.9).

For reducing the Silicon grains to use them in the solar cell for improved
efficiency, the Czochralski process (Fig. 2.10) was introduced to purify the Silicon.
It guarantees the uniform orientation of the Silicon grains, which increases the
efficiency of the solar cell. The Czochralski process has been named after Jan
Czochralski, a Polish scientist who introduced this process to the world in the
twentieth century. He started with melting polysilicon in a special sealed furnace
with a high temperature of up to 1500 °C. A well-oriented seed crystal (a base crystal
where the larger crystals will grow) was introduced to the molten Silicon to shape all

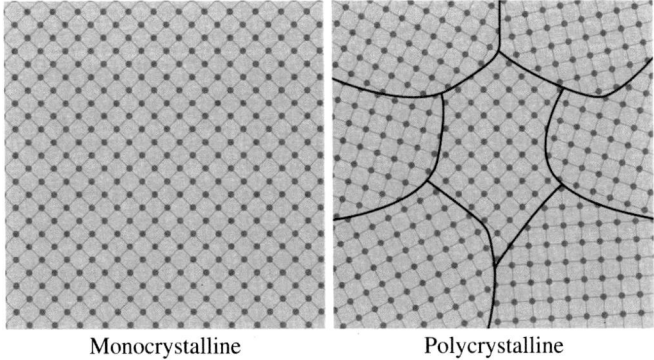

Monocrystalline   Polycrystalline

Fig. 2.9 Crystal orientation of (**a**) monocrystalline and (**b**) polycrystalline Silicon

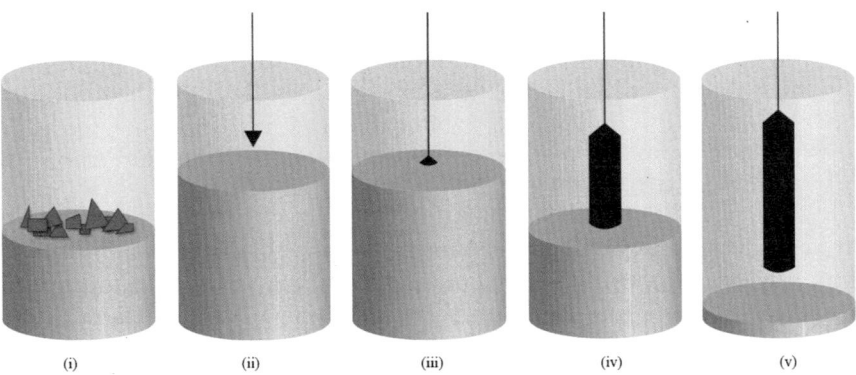

(i)   (ii)   (iii)   (iv)   (v)

Fig. 2.10 Consecutive steps of Czochralski process and scale of single crystal Silicon [4]. Step (i): Melting of Silicon and doping, Step (ii): Dipping the seed crystal, Step (iii): Beginning of crystal growth in top, Step (iv): Pulling the crystal, Step (v): Formed cylindrical shaped crystal with leftover melted Silicon

the molten Silicon in the same direction afterward. The orientation process is done by gradually cooling down the molten Silicon, while the seed crystal, which has the size of a pen, is pulled upward slowly. The molten Silicon starts to form a cylindrical shape over the seed crystals until all the Silicon is grabbed by the cylinder, which is full of well-oriented pure Silicon, ready to be used in any electronic or solar application. The size of the ingot can reach up to 2 m with a radius of 30 cm. It should be pointed out that the Czochralski process is not limited to a purifying process. Still, it can also be used for doping by adding positive or negative dopants, such as Boron or Phosphorous, respectively, in the melting phase. The schematic diagram of a Czochralski crystal puller is shown in Fig. 2.11.

Fig. 2.11 Schematic diagram of a Czochralski crystal puller [6]

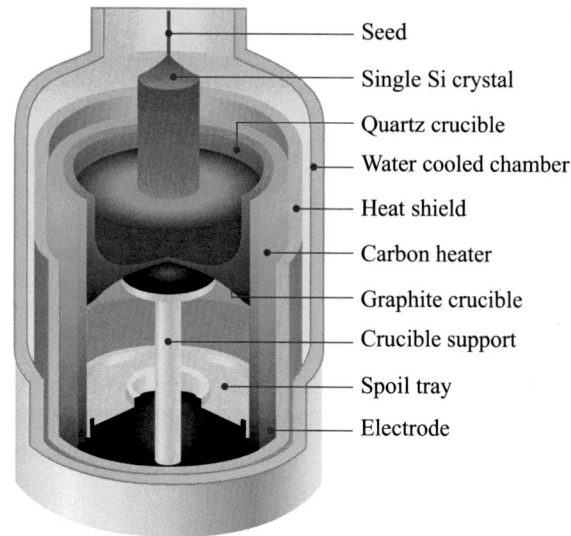

Seed

Single Si crystal

Quartz crucible

Water cooled chamber

Heat shield

Carbon heater

Graphite crucible

Crucible support

Spoil tray

Electrode

2.7 Wafer Formation

As a result of the ingot fabrication process, a long cylindrical monocrystalline ingot is obtained by the Czochralski process in addition to a parallel rectangular big polycrystalline ingot. These two ingots are cut into small wafers that have a variety of sizes and thicknesses based on the wafer technology. For instance, monocrystalline wafers sizes boosted from 125 × 125 mm in the first decade of the 2000s to 156 × 156 mm by 2015 with a thickness scale of 150–300 μm. The increase in ingot diameter from 165 to 200 mm is the cause of this expansion.

The monocrystalline ingot is cut into wafers using a sawing technique (Fig. 2.12), which consists of two to four rotating drums with several wires spaced apart, where the distance between them is the desired solar cell thickness. The ingot is oriented to pass through the high-speed rotating sharp wires, which play the role of a very sharp saw so that the robust ingot is cut into thousands of wafers. The obtained cells have a very rough surface and need an extra polishing process to guarantee a smooth surface area for further uses.

The polycrystalline ingot is cut into multiple sub-ingots by a huge saw where each sub-ingot is melted in a high-temperature resistive container. A resistive frame is also introduced into the molten Silicon and slowly pulled away to shape a polycrystalline layer (Fig. 2.13) on a substrate that can be removed after the Silicon is cooled down and solidified. This layer is cut into individual solar cells, which are ready for further chemical and physical processes toward being a perfect solar cell. This entire conversion process from polycrystalline ingot to the solar cell is referred to as the ribbon process.

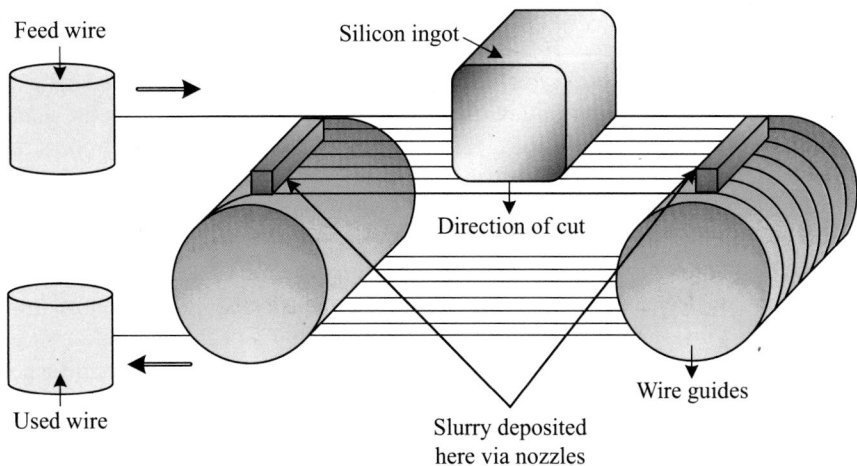

Fig. 2.12 Sawing technique to cut ingots into solar Silicon wafers

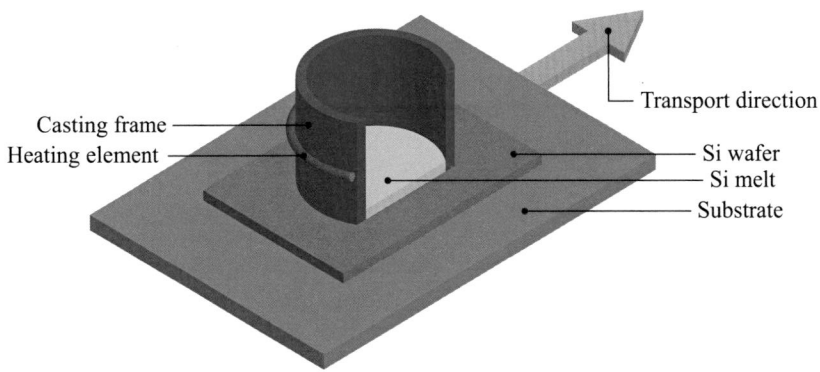

Fig. 2.13 Polycrystalline layer formation

The main advantage of the mentioned ribbon process is the absence of Silicon losses due to the thickness of the wires, also known as kerf loss. The average thickness of the wires is 100 μm, which is 50–66% of the solar cell thickness.

2.8   Cell Fabrication

A solar cell has a large area of a p-n junction. Solar cell formation starts with p-type Silicon that is obtained from the previously mentioned process, in which a p-doped ingot is formed and then cut into wafers. The non-uniformed and uneven surface of the wafers is cleaned up for the next process, which is called surface texturing.

2.8.1   Surface Texturing

The flat surface of the initial wafer has high reflectivity, which increases the optical losses of the solar cell by preventing some of the photons from penetrating the solar cell. Thus, crystalline Silicon solar cells must have a wafer texturing process to increase photon observation so that the electrons inside the p-n junction may be energized. The texturing process is done by a wet electrochemical etching process that forms a unique texture on the wafer surface to reduce the surface's reflectivity and trap the photons.

The most widely used etching solution in the industry is the solution of low concentrated Potassium hydroxide, water, and isopropyl alcohol. Several surface texturing methods are discussed in the literature to determine the best texture shape for the solar cell surface. A recent study has been done to compare two texturing methods with a clean and non-textured solar cell. The first texturing method creates porous Silicon (PS), while the other method creates pyramids Silicon [7].

If the root mean square of the high degree of roughness for the three cases—the as-grown Silicon layer, the PS texturing, and the pyramids texturing—are compared, it would be seen that the highest height is achieved by the PS texturing. The highest height implies that the cell is more likely to trap the photon and decrease sunlight reflection, unlike the as-grown Silicon case where the reflectivity reached 50%. This difference illustrates that 50% of the potential photons cannot reach the p-n junction without the texturing process, indicating a 50% optical loss. The percentage of reflected photons leads to optical losses, which are converted into electrical losses or reduction in solar cell efficiency. A study measured the ratio of the cell's electrical output to the input light illumination as the efficiency of the solar cell where the values of efficiencies were 13.23%, 11.36%, and 3.7% for porous Silicon, pyramids Silicon, and the original solar cell, respectively. Therefore, surface texturing is mandatory in all solar cell production to avoid unnecessary losses [8].

2.8.2   Adding the n-Type Layer

The solar wafer is now a textured p-type crystalline Silicon. However, p-type Silicon is not enough for achieving the PV effect; it needs an n-type material to be attached. Therefore, all wafers are placed to be fully exposed to a phosphoryl chloride gas that plays the role of the donor of the negative dopant. The mentioned process is done in a controlled furnace with an average temperature of 800–1000 °C. The phosphorous is deposited on the surface of the p-type material forming a p-n junction on each edge of the material. The thickness of the n-type material is controlled using the temperature and concentration of the dopant gas in addition to the period during which the wafers are exposed to the dopant gas.

During the process, an additional layer called phosphosilicate glass (PSG) is deposited on the p-type material. The PSG material is made of Phosphorus pentoxide (P_2O_5) and Silicon dioxide (SiO_2). This layer is undesirable and removed using a hydrofluoric acid wet etching.

To eliminate the junction at the edge of the wafers, they are piled on top of one another. The junction is then removed by etching the edges of the cells with extremely reactive plasma gas. Therefore, the p-n junction in the front is no longer connected to the rear one.

A technique known as plasma-enhanced chemical vapor deposition (PE-CVD) deposits a new layer on the n-type material. The layer is an anti-reflective coat made of Silicon nitride (Si_3N_4).

2.8.3 Screen Printing

Screen printing is used to provide a metal coating to the surfaces of the PV solar cells obtained from the previous steps. Screen printing was first developed in the 1970s, and it has become the most crucial part of solar cell fabrication. This process is mandatory to achieve the simplicity of the solar cells. The rear side of the wafer, which still has an n-type layer is screen printed along with a metal paste to create a reflective Aluminum layer. This layer plays a role in reflecting photons that might penetrate the wafer without any valuable contribution in creating electron–hole pair. The metal paste is pushed into the screen print mask through holes by dragging a squeegee across the screen. After removing the screen, a dense coating of wet metal paste is left behind. To remove the organic solvents and binders from the paste, the setup is dried in an oven. To fire the metal in contact with the Silicon, the wafer is further put in the second furnace of higher temperature. This firing process also plays a vital role in getting rid of undesirable n-layer on the rear side of the wafer. After the cleaning process, the wafer contains a metallic layer below a p-type material covered by an n-type material forming a p-n junction.

The obtained wafer after screen printing is flipped upside down for assembling the wafer in the PV module, where the n-type will be exposed to sunlight. Since n-type is not a perfect conductor, electric connectors are attached to the n-type in addition to the already discussed metallic layer in the bottom of the p-type to facilitate electron flow from p-junction to n-junction (inside the cell). This will facilitate the electron flow from p-type to n-type inside the cell and n-type to p-type outside the cell passing through the electric load.

The major steps from wafer formation to screen printing is summarized as follows:

Step 1: A rectangular p-type wafer is considered where an n-type layer is to be added above the surface. The surface is processed with an alkaline etch to smoothen the edges.

Step 2: The processed wafer is heated in a specialized furnace at a high temperature for adding Phosphorous around the wafer to form an n-type layer outside the p-type wafer.

Step 3: Several p-n layers are stacked on top to remove the junctions that formed at the edge of the wafer. A process called plasma etching is performed to break the junctions into smaller molecules which can be easily removed through vacuum systems. Therefore, the contact between the front and rear junctions is isolated.

Step 4: An Aluminum metal paste is added at the rear side of the cell, with a secondary silver paste. This metal paste helps to reduce the recombination of carriers with the surface, by creating a field called back surface field (BSF). High p-type doping at the rear cell also helps to increase the BSF and keep the electrons away from the end for obtaining increased cell voltage. There are some remaining materials on the wafer after applying the metal paste, which is removed by putting the setup in an oven.

Step 5: The cell is put into a second furnace to connect the metal contact with the Silicon at the rear of the cell. As a result of the temperature rise, the n-type material is removed from the rear side, and the solidified metal paste connects with the p-type layer.

Step 6: The cell is flipped for putting the metal paste at the front side of the cell.

Step 7: A similar metal paste is applied at the front side with an alternating pattern to avoid the shading effect. By putting the setup in a furnace, the contacts are connected into the Silicon. After this process, the cell can be fabricated into solar modules for manufacturing.

The mentioned steps of the development of a typical solar cell are depicted in Fig. 2.14.

The screen printing of solar cells has a significant disadvantage of shading due to the metallic contact on the n-type layer. This layer prevents the solar cell from being fully exposed to the sunlight, which means a lesser effective area on the solar cell surface. Therefore, the burial of metallic contact within a groove in the solar cell is introduced to develop a high-efficiency commercial solar cell technology. The main concept of this buried-contact technology is to maintain the same metallic contact volume while increasing the n-type layer's electrical conductivity but with fewer shades by sinking the metallic contact in the solar cell. For this reason, the metallic contact on the surface will cover less surface area. In other words, a wide surface metallic contact is replaced with a strip of metal in the solar cell.

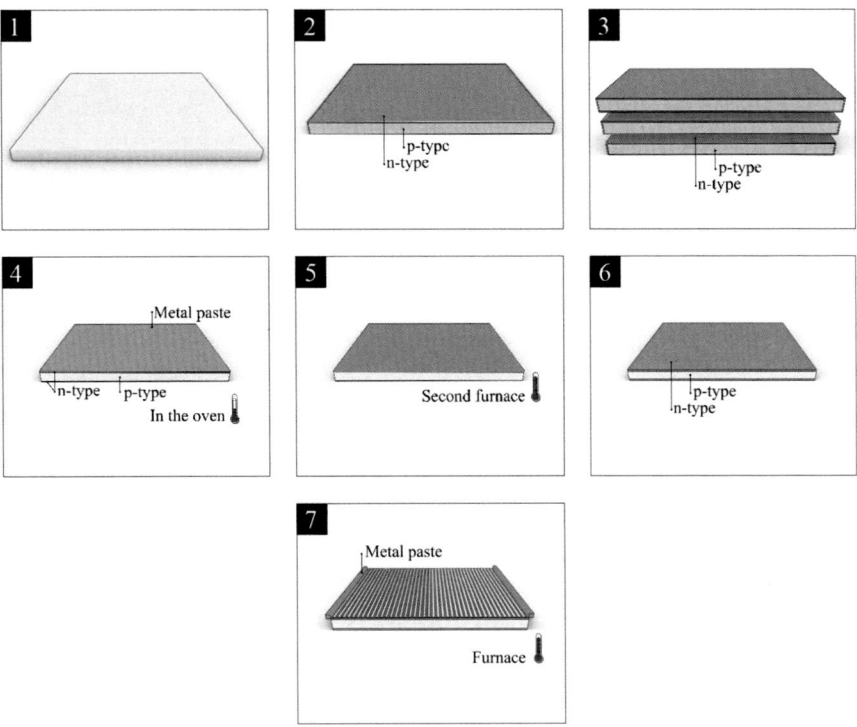

Fig. 2.14 Schematic diagram of a screen-printed solar cell

The same surface texturing and n-type deposition process is repeated as they were done in the screen-printed cells. Now, the cells are ready to create grooves using a laser or a fine-controlled mechanical saw that digs about 80 μm with a width of 30 μm. Due to the laser or saw cut, the n-type material does not cover the groove surface area of the p-type material. Another second heavy phosphorous doping is further used to mask the groove surface area with the n-type layer.

A layer of Aluminum is later added to the solar cell's rear part by screen printing or gas evaporation. A very high temperature is applied to the wafer for a long time to ensure the melting of Aluminum in the Silicon. The n-type layer attached to the bottom of the p-type material is completely removed.

A highly conductive layer of Copper starts filling the groove spaces in addition to covering the bottom part of the p-type material. The sides of the solar cell are cut to separate n-type and p-type layers and avoid a short circuit. The cell is now ready to be connected with other cells to form the solar module. Figure 2.15 demonstrates the steps of making buried-contact solar cells.

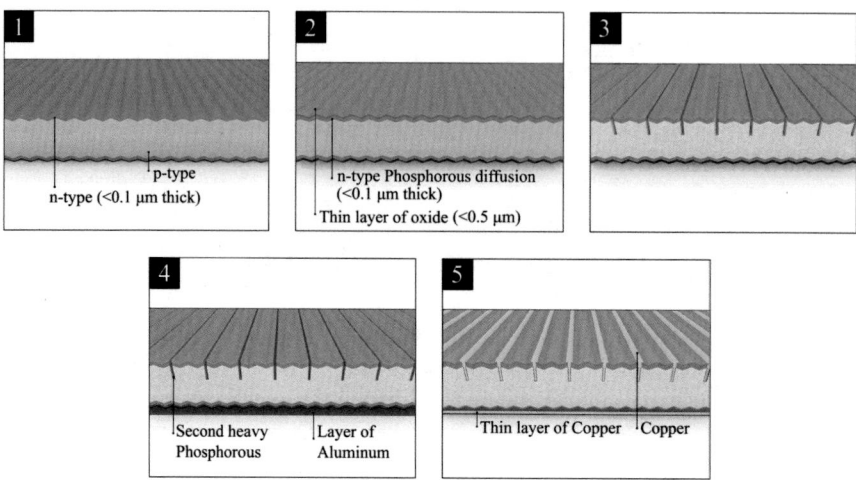

Fig. 2.15 Schematic diagram of a buried-contact solar cell

2.9   PV Module Fabrication and Construction

To form the solar module, which essentially contains many solar cells, one or multiple metallic strips called busbars are connected to the surface of the solar cell. The busbars collect the entire flow of electrons or electrical current from the n-type metallic layer and deliver it to the next solar cell until it passes through the electric load. The size and number of the busbars are critical to be determined due to their shading effect on the solar cell. Moreover, they have an internal resistance that leads to voltage drops or losses. Therefore, researchers are working on minimizing the overall equivalent resistance and the passive area underneath the busbars. Several studies are being performed to determine the best busbar numbers and size to achieve minimum energy losses or maximum efficiency.

The shading factor is the ratio of the shaded part of a PV module with respect to the total area of the module. This factor ranges from 0 to 1, where 0 refers to the condition with no shading and 1 refers to the condition with complete shading of the module. A study on the relationship between the number of busbars with the shading factor in addition to the effect of the busbar width on the shading factor, which directly influences the efficiency of the cell, shows that the change is linear. The shading factor is linear with the number of busbars, which varies between 4% and 9% when the number of the busbar is from 1 to 6. Likewise, any increase in the busbar width will lead to an increase in the shading factor. Results show that when the busbar width varies from 1–6 mm, the shading factor varies from 4 to 12% [9].

Experiments showed a dramatic increase in efficiency from 17.4% to 19.3% when the number of busbars increased from 1 to 2. Moreover, moving toward 3 or 4 busbars will increase the efficiency, but it will not have the same dramatic effect

Fig. 2.16 Four-busbar PV
cell

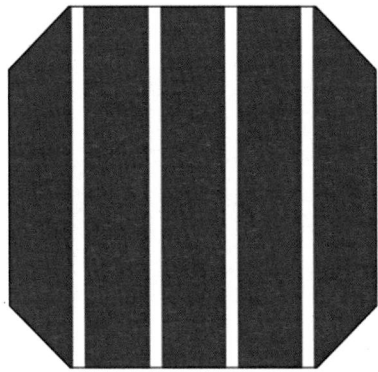

on the efficiency. Studies concluded that thin metallic busbars show outstanding performance toward increasing efficiency, and 4 busbars (Fig. 2.16) are the most economical number for conventional solar modules [9].

2.9.1 Encapsulant

Encapsulant plays a vital role in protecting solar modules from dirt and water. The encapsulant is a transparent, low thermal resistive material that covers the solar cells from both sides like a sandwich. Solar cells are encapsulated before covering the cells with solar module glasses. The most commonly used encapsulant is ethyl vinyl acetate (EVA). Some other encapsulants are Targray, Tedlar, etc. Encapsulation of solar modules ensures the better durability and performance of the modules.

Once the cells are assembled and electrically connected using the metallic busbars, they form the module's core as an array of cells. This array is centered on a layer of transparent encapsulant material that is attached to a glass layer. The transparent encapsulant material such as EVA is used, which is a thermoplastic polymer that gets melted and takes the shape of the array when heated to about 120 °C. Another encapsulant layer is inserted in the rear side of the cell, so the solar cell array is sandwiched between two encapsulants.

2.9.2 Glass and Backsheet Layers

Glass is the top layer of a solar module, which gives it mechanical stability and protection against water and dirt. Solar module glasses are made of tempered glasses, which are strong enough to provide safety to the solar modules. This layer must be frequently cleaned to avoid dirt and soil deposition, which prevents the photons from reaching the cell and causes losses called soiling losses. Glass might also cover the rear side of the cell in some state-of-the-art modules called bifacial solar modules.

Fig. 2.17 Components used in solar module encapsulation. The layer of solar cell is covered with two different layers at both the sides

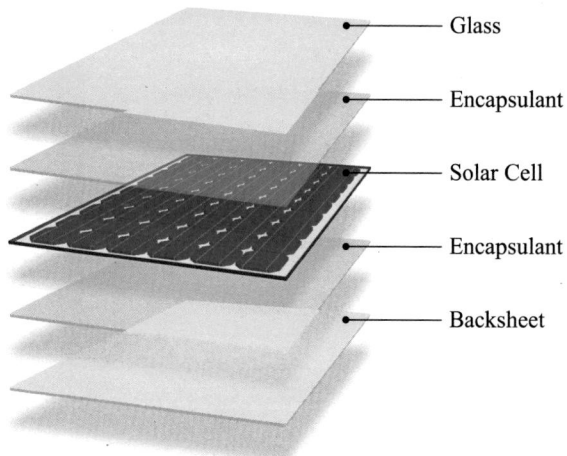

Glass

Encapsulant

Solar Cell

Encapsulant

Backsheet

After the solar cells are sandwiched with encapsulants, a backsheet layer is also added to the rear side of the array. The backsheet layer protects the module from humidity and other mechanical stresses. The lamination process is highly critical due to the temperature and pressure applied to the cell, in addition to the necessity of avoiding air bubbles during the process, which would otherwise be considered a failure for any produced module. The components for encapsulating the solar module are shown in Fig. 2.17

2.9.3 Frame

A frame gives the module the ability to be attached and mounted on the mechanical structure. The frame is usually made of aluminum due to its corrosion resistance, strength, and lightweight. To ensure the safety along with the proper operation of the solar modules, all modules' frames are electrically connected, and then grounded.

Afterward, all the electrical contacts are combined to a junction box placed at the top of the rear side of the module. The junction box contains a few diodes called bypass diodes that play a significant role which will be discussed later. Even though the module is electrically functional after connecting the junction box, it still needs mechanical support and a mechanical frame to mount it on a roof or any ground mounting structure. Thus, an Aluminum frame was introduced to put around the module [10].

2.9.4 PV Module Label

PV module label is a concise datasheet that includes the most important characteristics of the module. This contains manufacturer information, maximum power point (MPP), MPP voltage, MPP current, open-circuit voltage, short-circuit current, NOCT, maximum system voltage, operating temperature, weight, dimensions, and many other information about the module. In short, it contains power rating, output characteristics, and the details of the manufacturer. A solar module has to go through specific tests before getting into the commercial zone. The icons of certification of these tests are also provided on the bottom part of the label.

2.10   Summary of the Steps of Making a PV Module

The steps of making a PV module are summarized in Fig. 2.18. The steps are described below [11]:

Step 1: The automated glass loading is prepared as the base of the PV module.

Step 2: The first encapsulated sheet is laid down on the glass loading shown in Step 1.

Step 3: The PV cells are loaded, tabbed, and finally connected in a string through contactless infrared soldering on cells.

Step 4: The strings from Step 3 are placed on the glass and encapsulant slab in Step 2. The string connector ribbons are manually or automatically soldered.

Step 5: The second encapsulated sheet is laid down.

Step 6: The backsheet is laid down. Visual inspection and electroluminescence testing are done. The module is laminated.

Step 7: The edges of the module (shown in black) are trimmed in an automated way. An Aluminum frame and Silicone sealant are added.

Step 8: The string connector leads are soldered in the junction box and filled with potting agent. The sticker is placed and the module is cured.

Step 9: The I-V module is tested and electroluminescence inspection is done. The module is sorted and palletized.

After these nine steps, the module is ready for shipping to the warehouse, from where it is shipped to customers.

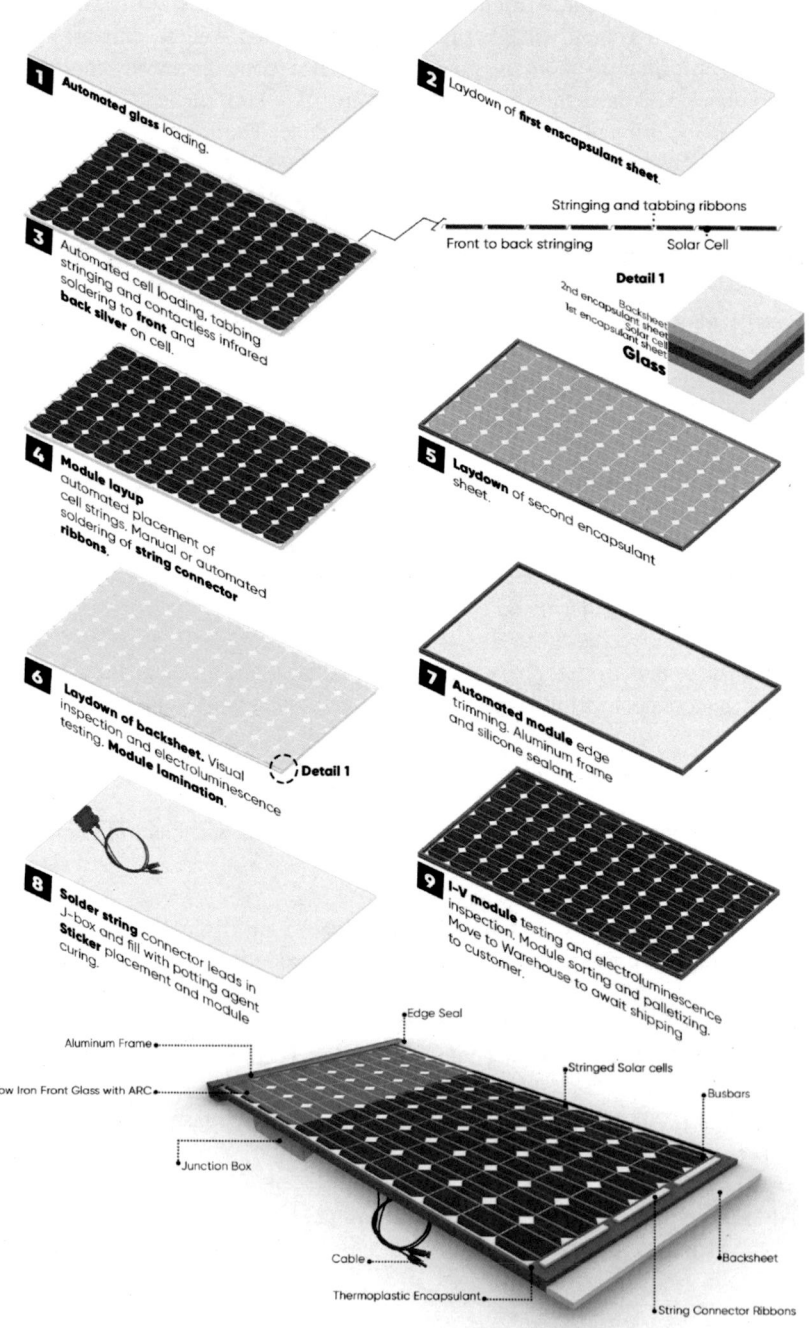

1 Automated glass loading.

2 Laydown of **first encapsulant sheet**.

Stringing and tabbing ribbons

Front to back stringing   Solar Cell

Detail 1

2nd encapsulant sheet
Backsheet
Solar cell
1st encapsulant sheet
Glass

3 Automated cell loading, tabbing stringing and contactless infrared soldering to **front** and **back sliver** on cell.

4 **Module layup** automated placement of cell strings. Manual or automated soldering of **string connector ribbons**.

5 **Laydown** of second encapsulant sheet.

6 **Laydown of backsheet.** Visual inspection and electroluminescence testing. **Module lamination**.

Detail 1

7 **Automated module** edge trimming. Aluminum frame and silicone sealant.

8 **Solder string** connector leads in J-box and fill with potting agent **Sticker** placement and module curing.

9 **I-V module** testing and electroluminescence inspection. Module sorting and palletizing. Move to Warehouse to await shipping to customer.

Edge Seal

Aluminum Frame

Stringed Solar cells

Low Iron Front Glass with ARC

Busbars

Junction Box

Cable

Backsheet

Thermoplastic Encapsulant

String Connector Ribbons

Fig. 2.18 Steps of making a PV module [11]

2.11   Packing Density

Any unused space on a solar PV module makes the module inefficient, as it does not play a role in converting light into electricity. So, a dense packing density is desired in PV modules. Packing density basically refers to the total surface area occupied by the solar cells with respect to the part which is vacant. The packing density varies based on solar cell technology because of the varieties found in the shape of the solar cells in each fabrication process.

As discussed in the previous sections of the chapter, monocrystalline cells are cut from the cylindrical Czochralski process using a saw. Therefore, the produced cells will have a circular shape, while the polycrystalline Silicon solar cells are extracted from rectangular ingots and have a square shape.

As illustrated in Fig. 2.19, the packing density of the monocrystalline module is less than the polycrystalline one, which means a decrease in the efficiency since the spaces between the round cells are not effective and will not participate in the energy-yielding process. Solar module manufacturers began a new approach to maximize the packing density of monocrystalline modules in 2005 by trimming the ingot into a hexagonal shape (Fig. 2.20) after cutting the head and the tail of the original cylindrical ingot. The final hexagonal-shaped solar cell has less surface area compared to the round one. However, the round modules' overall cost and efficiency losses are reduced by having a 156.75 mm × 156.75 mm flat-to-flat cross-section area out of 244 cm^2 area instead of the 314 cm^2 that is obtained from the round solar cells [10].

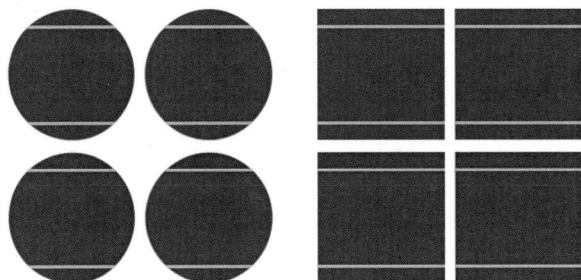

Fig. 2.19 Packing density difference between monocrystalline (left) and polycrystalline (right) modules

Fig. 2.20 Maximizing the packing density in monocrystalline modules

2.12 Cells Connected in Series

Ever since the solar modules entered the commercial industries, they have been tightly attached to the battery industry, which has been the most common energy storage technology as DC output power is obtained from the solar modules. Thus, the early solar modules were designed to meet the 12 V battery charging process standards. Each Silicon-based solar cell has almost 0.5–0.68 V capability under the standard test conditions. Therefore scientists decided that connecting 36 cells in series is a good option to charge a 12 V lead-acid battery in any situation. But the voltage of the solar cell might go significantly down when the ambient temperature gets high, which will be described in detail in Chap. 3.

The series connection is made using the busbars, which connect the negative front side of the cell to its positive rear side. The series connection will give the solar module 18 V between the positive and negative terminals, where the value is obtained from Eq. 2.5.

$$Potential\ difference\ between\ two\ terminals$$
$$= Number\ of\ solar\ cells \times Each\ solar\ cell\ voltage. \qquad (2.5)$$

In a series connection, to obtain 18 V across the negative and positive terminals, 36 solar cells are connected where each of them has a voltage of 0.5 V. Moreover, this connection will give equal current to the solar cells as the series configuration

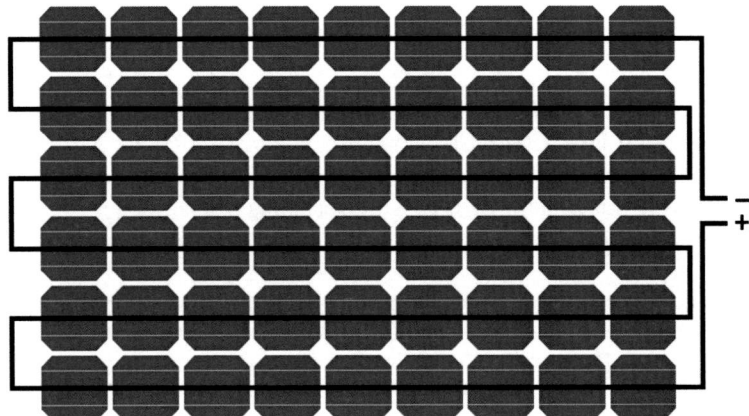

Fig. 2.21 A series connection between the PV cells

keeps the current fixed. Figure 2.21 illustrates the series connection of solar PV cells. Nowadays, the modules usually consist of a series connection of 36, 60, or 72 Silicon-based solar cells. In fact, bifacial modules and half-cut cells have made it possible to incorporate 120 or 144 cells in each module.

2.13 Hotspot Heating

Hotspots appear when one or multiple solar cells connected in series are shaded by snow, lead, dirt, or hard shadows. The shaded solar cell(s) will not be able to generate the same current value as the neighboring illuminated cells. The shaded cell will limit the current to the value that it generates, causing a lot of energy loss or dissipation in the form of heat. Since the heat resulted by the energy dissipation takes place in a relatively limited and small area, the temperature rises near the defective zone starting from 1–80 °C and can increase more than 200 °C, causing one or several effects such as low damage, backsheet bubbles, glass cracking, or cell cracking when the temperature reaches more than 200 °C.

2.13.1 Causes of Hotspots

Hotspot phenomena might show up or increase due to several causes, such as:

1. **Cell mismatch:** Series connection of solar cells having dissimilar current ratings can cause more energy dissipation as heat, which increases the temperature at some places of the cell, causing hotspots.
2. **Shading of solar cell:** Partial shading in any solar cell or any string of cells can be a major disadvantage in the solar cell, causing high reverse-biased current in

the shaded part. This increases more heat dissipation on the shaded solar cell, and thus hotspot is seen.

3. **Manufacturing defects:** The manufacturing of solar cells includes many steps, and fault in any step can cause a hotspot when the solar module is in use.
4. **Contaminating impurities:** Contaminating impurities such as dirt, sand, etc. give rise to hotspots.
5. **Solar module installment defect:** While installing a solar module, it is necessary to check the roof conditions to avoid unnecessary shadings; otherwise, it causes hotspots.
6. **Imperfect screen printing:** Screen printing of solar cells includes many steps such as adding an n-type layer, isolating edges, and many more. Defects caused in these stages can disrupt the flow of electrons and electricity in between the two terminals, further causing hotspots.

2.13.2 Effects of Hotspots

Hotspots can have several detrimental effects on the solar PV system, which includes:

1. Increase of the equivalent resistance of the solar cell that disrupts the current flow.
2. Rise in temperature that may lead to a fire.
3. Rise in temperature that can melt the solar cells at small places, making holes.
4. Breakage and melting at many small places of the solar module can cause shattering.

All these defects should be detected using several methodologies before selling the module. However, some defective modules may still be available in the market and installed in a large utility-scale solar plant. So, how will the defective cell or module be detected? Nowadays, thermal cameras fixed on a drone take video shots to detect the hotspots before delivering the project to the end customers. The defective module can be changed, and more improved performance can be ensured [10].

2.14 Nominal Operating Cell Temperature

All the electrical characteristics of the solar modules are taken under specific standard test conditions, which may differ from the actual operating parameters. Solar modules are installed worldwide with different ambient temperatures, tilt angles, Sun positions, and solar irradiance. Due to the direct effect of these factors, it is essential to know the actual operating conditions such as the cell operating temperature, which is the actual temperature of the solar cell when the PV module is operating in some specific conditions, such as:

- Solar radiation: 800 W/m^2.
- Ambient temperature: 20 °C.
- Wind speed: 1 m/s.

The *nominal operating cell temperature (NOCT)* is defined as the temperature of open-circuited PV cells subject to the above conditions. Most manufacturers consider the value of NOCT as 41–46 °C.

The ambient temperature can be obtained from the weather data. The cell temperature from the available ambient temperature can be determined using the formula in Eq. 2.6.

$$T_{solar\ cell} = T_{ambient} + \frac{NOCT - 20}{80} \times S, \qquad (2.6)$$

where S is the irradiance in mW/cm^2.

An example of the determination of the PV cell temperature from the NOCT and solar irradiance is given below

Example 2.1 (Determination of PV Cell Temperature from NOCT and Solar Irradiance) A solar module is installed in Cairo, Egypt, where the average daytime ambient temperature is 35 °C in June, July, and August. What is the average temperature of the PV cell when the irradiance is 80 mW/cm^2, and the NOCT is 46 °C?

Solution

- Ambient temperature, $T_{ambient} = 35$ °C,
- Irradiance, $S = 80$ mW/cm^2.
- According to Eq. 2.6,

$$T_{solar cell} = 35 + \frac{46 - 20}{80} \times 80 = 61\ °C.$$

So, the operating temperature of the solar cell is greater than the ambient temperature by $61 - 35 = 26$ °C while the PV module is installed on an open backside mounting system.

However, the PV modules can be installed on different mounting systems that might block the cooling process provided by the wind, for instance, in a rooftop mounting system, where the module temperature is expected to be much higher than the open backside of the module. Therefore, a new terminology called *installed nominal operating cell temperature (INOCT)* has been introduced to consider the effect of the mounting configuration type on the solar module's actual temperature.

Table 2.1 Values of X based on the standoff/entry/exit height (whichever is minimum) [12]

Height	X
1 in.	11
3 in.	2
6 in.	−1
9 in.	−3
If there is channeling under the array	4

For rack mounting, the INOCT is 3° less than the NOCT. For direct mounting, the INOCT is 18° more than the NOCT. However, for standoff or integral mounting, the relation between INOCT and NOCT is determined using the formula shown in Eq. 2.7.

$$INOCT = NOCT + X, \qquad (2.7)$$

where X is the additional temperature in °C depending on the mounting type, which is determined from Table 2.1. The standoff height can be measured from the roof to the frame of the module.

2.15 Testing and Device Reliability

Almost all solar module manufacturers provide a 25-year linear warranty for their crystalline Silicon-based solar modules. This warranty guarantees that the module's overall performance will not be less than 80% until 25 years of the production date. In other words, the overall degradation and deficiencies will not exceed 20%; otherwise, the manufacturer is supposed to refund or replace the module based on the given warranty.

During the relatively long lifetime, the module will face some rough atmospheric conditions and various external stresses such as mechanical and electrical ones. Crystalline Silicon solar modules have a series of standard accelerated testing procedures provided by the International Electro-technical Commission (IEC) under the section of IEC 61215. The accelerated testing steps can be illustrated as the following:

1. **Visual inspection:** The main goal of this test is to check any major defects visually or cracked, broken cells, substrate, junction box, or even the Aluminum frame. It is also used to detect any air bubbles between the cells and the encapsulants and double-check that the backside label is still attached to the module. This test is frequently conducted throughout the other tests to assure that the other testing procedures do not cause any defect.
2. **Maximum power:** This test is performed by exposing the module to the sunlight or using a simulator in the laboratory. Most laboratories have a Sun

simulator to simulate the *standard test conditions (STC)*, which correspond to:

- Air Mass Index (AMI) of 1.5.
- Cell temperature of 25 °C.
- Light or solar radiation of 1000 W/m^2.

This test is essential to compare the theoretical maximum power that the module should provide with the actual maximum power that is practically delivered.

3. **Insulation resistance:** This test ensures that the current-carrying components of the module are adequately insulated from the Aluminum frame. It is done by applying a DC voltage from an external source with a value up to 1000 V and twice the maximum voltage of the system. The module passes the test if there is no failure or any surface tracking. In the large modules with a surface area greater than 1000 cm^2, the resistance should be greater than 40 MΩ for every 1 m^2 area.

4. **Wet leakage current test:** It is listed under the electrical and safety test. The main purpose of the test is to check that the module is insulated against moisture penetration when installed in wet operating conditions such as rain, fog, and melted snow. The module must prevent any water leakage to avoid corrosion reactions inside the module. The tester submerses the module in a shallow water tank where the module is fully surrounded by water except for cable entries of the junction box. A test voltage up to the maximum module voltage is provided between the shorted output terminals and the water bath solution for 2 min. The test detects any failure in the design of the connector.

5. **Temperature coefficients:** The temperature affects all the electrical parameters of the solar modules, including output voltage, current, and power. So, it is vital to determine the short-circuit current, power coefficients, and open-circuit voltage due to their important role in the design process for all PV designers around the world where the temperature varies. The test is done under constant irradiance with different temperatures during the heat-up process or cool-down process. The simulator takes the I-V curve at every 5 °C and derives the coefficients afterward. The I-V curve will be discussed in detail in Chap. 3.

6. **NOCT:** As discussed earlier, the NOCT is an essential factor provided by the manufacturer to help the designer in the calculation process of the actual temperature of the module. The test requires the following standards:

- Total irradiance: 800 W/m^2.
- Tilt angle: 45°.
- Wind speed: 1 m/s.
- Ambient temperature: 20 °C.
- No electrical load is connected to the module's terminals.

The test has a data logger connected to a temperature sensor connected to the backside of the module, ambient temperature sensor, and wind speed sensor to obtain the NOCT.

7. **Outdoor exposure:** This test measures the module's ability to withstand the outdoor temperatures throughout its lifetime by exposing the module to an irradiance of $60\,kWh/m^2$. Although the test period is short, it might be a good indicator of any future defects or failures resulted from outdoor conditions.

8. **Hotspot endurance:** It determines whether the solar modules can endure the heat that occurs in the module due to partial shading, cracked cell, or even cell mismatching.

9. **Bypass diode:** The bypass diodes are electronic components that are put in a junction box and attached to the upper backside of the module. The test is done by measuring the temperature of the diodes when the module is heated up to $(75 \pm 5)\,°C$ and applying a short-circuit current as listed in the datasheet for an hour, and then the current is raised to 1.25 times of the short-circuit current for an hour. The bypass diodes and their soldered connections must withstand the two testing cases; the backsheet and encapsulant are supposed to hold the high temperature. Apart from that, the diodes must be oversized.

10. **Ultraviolet (UV) preconditioning:** This test is done by exposing the solar module to UV irradiance of $15\,kWh/m^2$ in the range of (280–400 nm) while the module's temperature is kept at $60\,°C$. The main goal is to ensure that the solar module is resistant to any defect or breakdown caused by the UV irradiance, which usually leads to discoloring of the encapsulant and the backsheet to yellow. This yellowing phenomenon decreases the efficiency of the module by preventing enough light from penetrating the cell.

11. **Thermal cycling TC200 (200 cycles):** This is a very critical test through which the manufacturer applies thermal stress on the module and monitors the behavior of the soldered connections in the laminate. Since the connectors are metallic materials, they have different expansion coefficients, which might lead to defects inside the module. The test is done in the range of $(-40 \pm 2)\,°C$ and $(+85 \pm 2)\,°C$, and the failure rate can reach 30–40%.

12. **Humidity-freeze:** It detects the ability of the junction box to show enough resistivity toward delamination or adhesion by applying a very high temperature combined with humidity then followed by a very low temperature.

13. **Robustness of terminations:** The test is done to measure the robustness of the termination of the solar module, such as the flying leads and screws. It is listed under mechanical stress tests.

14. **Damp-heat DH1000 (1000 h):** Along 1000 h, the solar module is exposed to $(85 \pm 5)\%$ humidity with an applied temperature of $(85 \pm 2)\,°C$. The module and its junction box are supposed to withstand these circumstances, and no adhesive loss and elasticity loss of the encapsulant or the junction box are supposed to occur.

15. **Mechanical load test:** It is a simulation of the mechanical load that the module is expected to face, such as ice, snow, wind, and static load.

16. **Hail impact:** The solar modules are supposed to withstand the impact of ice balls with a diameter of 25 mm that fall at a specific falling velocity (23 m/s) [13].

2.16 Degradation and Failure Modes

During manufacturing the solar cells or forming, testing, and installing solar modules, modules, in general, might be defected by reasons related to a failure due to human or machine error or an environmental reason associated with the geographical location of the solar station. These defects can be summarized as the following:

1. **Module Delamination:** It is a systematic degradation for all PV modules that happens when adhesion between the front glass, the transparent encapsulant, and the backsheet occurs. It is caused due to the weak bonds between the components used in PV cells due to some environmental operating conditions such as moisture, thermal expansion, or aging.
2. **Junction box failure:** These failures refer to the poor fixing materials of the junction box placed on the PV module's backside or the poor installation of the junction box. Junction box failure causes water or moisture penetration to the electronic components inside the junction box, leading to corrosion in the metallic connections between the bypass diode and the terminals of the module. Moreover, bad bypass diodes soldering is the main reason for arc failure in the junction box, igniting a fire.
3. **Module glass breakage:** It is mostly caused by a mechanical force such as snow load, ice, extreme thermal stress, wind, and hail drops. In some conflict zones, shrapnel might be most likely because of a module glass failure.
4. **EVA discoloration:** EVA is one of the encapsulants used in the PV modules, which might lose its transparency due to discoloration resulted from environmental and aging conditions. UV radiation is also a main cause of browning or yellowing the EVA layer, which causes a major decrease in the module output.
5. **Cell cracks:** The thickness of the solar cell is 150–200 μm, which makes it easier to be cracked with minimal mechanical force. These cracks might be visible to the naked eyes or even needs special techniques to locate the cracked cell in a module. The damage is a possible failure during the processing and assembling of the solar module, which is not inspected instantly but later on. Cracks can start from the cell itself when the manufacturer starts connecting the ribbons (busbars) on the cell during the soldering process or starting from the edge of the cell when a hard object bounces the cell.
6. **Burn marks:** The main cause of this failure is the initial failure of the soldering bonds, ribbon breakage, or localized heating. All these failures are physically converted to resistance which increases toward the passing current through the cell, which will lead to extreme heat followed by unhandled thermal stress on the cell. These marks can also be a result of a failure in the bypass diodes, which

leads to reverse bias on the cell, which will cause an electrical pressure on it with a burn mark that can be visually inspected [14].

2.17 Conclusion

This chapter discusses the manufacturing and assembling of solar PV cells, followed by different tests to consider the formation of a more reliable solar module. The formation of the p-n junction and necessary clarification of n-type and p-type Silicon has been discussed in this chapter. Extraction of pure Silicon through different reactions along with figures has been discussed, which goes through the Czochralski process for more purification and growing ingots. Doped Silicon has to go through many processes to take the structure of an ideal solar cell in terms of shape and quality, which have been manifested in this chapter. Even though the manufactured solar cells go through many tests to verify their reliability, they face some degradation with time because of many reasons pointed out in this chapter. This chapter will assist the readers in gaining knowledge about all the steps that need to be followed for making a solar cell.

2.18 Exercise

1. Explain the differences between n-type Silicon and p-type Silicon.
2. What is meant by surface texturing?
3. What is hotspot heating? Explain the drawbacks of having hotspots in a solar cell.
4. Briefly narrate the steps of screen printing.
5. Explain the role of NOCT in manufacturing solar cells. What are the standards required to be followed while testing the device's reliability?
6. How many solar cells are needed to be connected in series to get an overall potential difference of 30 V across the terminals of the module under standard test conditions?
7. Why is surface texturing of solar cells essential?
8. In your opinion, how can hotspot heating be avoided?
9. If the operating solar cell temperature is 50 °C at 29 °C ambient temperature, find the value of NOCT when the solar irradiance is 86 mW/cm^2. [Answer: 39.535 °C]
10. What is the difference between monocrystalline and polycrystalline Silicon crystals? Explain with figures.

References

1. R.E. Hummel, *Electronic Properties of Materials* (Springer, Berlin, 2010)
2. L.S. Bobrow, *Fundamentals of Electrical Engineering*, chapter 6 (Oxford University Press, Oxford, 2003)
3. Semiconductor Technology, Fundamentals: doping: n- and p-semiconductors (2021). https://www.halbleiter.org/en/fundamentals/doping/
4. M.R. Seacrist, G. Fisher, R.W. Standley, Silicon crystal growth and wafer technologies. **100**, 1454–1474 (2012)
5. Z. Liu, W. Ma, X. Ye, Chapter 2—shape control in the synthesis of colloidal semiconductor nanocrystals, in *Anisotropic Particle Assemblies*, ed. by N. Wu, D. Lee, A. Striolo (Elsevier, Amsterdam, 2018), pp. 37–54
6. S. Meroli, Czochralski process vs float zone: two growth techniques for mono-crystalline silicon (2021). https://meroli.web.cern.ch/lecture_silicon_floatzone_czochralski.html
7. J. Escarré, D. Ibarz, S. Martín de Nicolás, C. Voz, J.M. Asensi, J. Bertomeu, D. Muñoz, P. Carreras, Optimization of KOH etching process to obtain textured substrates suitable for heterojunction solar cells fabricated by HWCVD. Thin Solid Films **517**(12), 3578–3580
8. K.A. Salman, Effect of surface texturing processes on the performance of crystalline silicon solar cell. Solar Energy **147**, 228–231 (2017)
9. A. Qadir, A. Qayoom, Q. Ali, The effects of metallization of busbars on the performance of PV cell. Clean Energy Technol. **7**(1), 7–10 (2019)
10. M. Woodhouse, B. Smith, A. Ramdas, R. Margolis, Crystalline silicon photovoltaic module manufacturing costs and sustainable pricing: 1h 2018 benchmark and cost reduction road map (2020)
11. A. Ramdas, M. Woodhouse, B. Smith and R. Margolis, Crystalline silicon photovoltaic module manufacturing costs and sustainable pricing: 1h 2018 benchmark and cost reduction roadmap. co: National renewable energy laboratory (2021). https://www.nrel.gov/docs/fy19osti/72134.pdf
12. M.K Fuentes, A simplified thermal model for flat-plate photovoltaic arrays. Technical report, Sandia National Labs., Albuquerque, NM (USA), 1987
13. R. Arndt, I.R. Puto, Basic understanding of IEC standard testing for photovoltaic panels. https://eif-wiki.feit.uts.edu.au/_media/solar_panel_iec_standards_testing.pdf
14. M. Köntges, S. Kurtz, C. Packard, U. Jahn, K.A. Berger, K. Kato, T. Friesen, H. Liu, M. Van Iseghem, Review of failures of photovoltaic modules (2014). https://iea-pvps.org/wp-content/uploads/2020/01/IEA-PVPS_T13-01_2014_Review_of_Failures_of_Photovoltaic_Modules_Final.pdf

Solar Cell Properties and Design

3

3.1   The Effect of Light on Solar Cells

A Silicon-based solar cell is a p-n junction formed by the integration of n-type and p-type silicon layers. A p-n junction has two terminals with a potential barrier, where one terminal is the anode, and the other is the cathode. It allows the current to flow in one direction while blocking the reverse flow like a diode. This electrical device converts light energy into electrical energy, whose electrical characteristics (voltage, current, and resistance) depend on the light that falls on the solar cells.

Though solar cells are verified using the standard level of irradiance, the actual output found by sunlight exposure is quite different from the expected output. The output obtained from the solar modules depends on the intensity of irradiance, and as it is not always consistent, the output varies. This variation also depends on the type of materials in the solar cells. Studies say that the decrease in current in solar modules is linear and voltage is logarithmic concerning the decrease in light availability. This decrease is mainly caused by the effect of shunt resistance, as the equivalent resistance matches up with the shunt resistance with the decrease of the light level. Thus, a large amount of total current passes through the shunt resistance, increasing the power loss. The amount of loss depends on the materials comprising the solar cells and the manufacturing process [1].

As the solar cells are mainly formed of diodes, this chapter will familiarize the readers with the concept of diodes and their functioning. The ideal diode is a bipolar electronic device where the current passes in one direction when a forward threshold voltage is applied. In contrast, it prevents any current from flowing in the opposite direction. Figure 3.1 illustrates the basic representation of a diode. However, in practice, when the reverse voltage is applied, diodes always have some minimal leakage current, called dark or reverse saturation current.

© The Author(s), under exclusive license to Springer Nature Switzerland AG 2022   57
Y. Abou Jieb, E. Hossain, *Photovoltaic Systems*,
https://doi.org/10.1007/978-3-030-89780-2_3

Fig. 3.1 Basic
representation of a diode

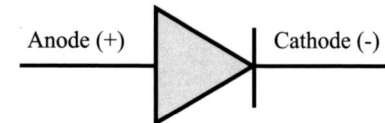

Fig. 3.2 The behavior of p-n
junction in dark and under
illumination

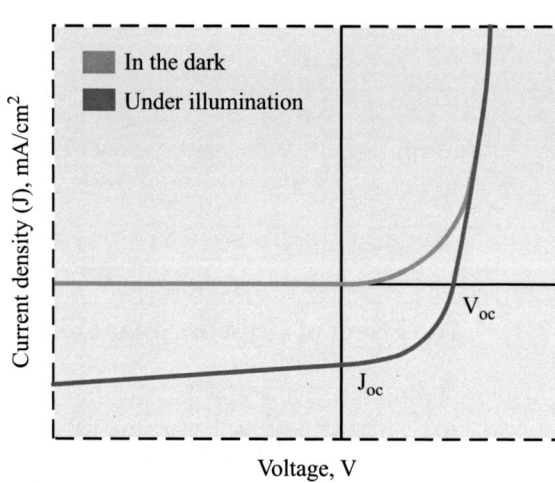

Voltage, V

The current passing through a diode in the forward direction can be represented
by Eq. 3.1:

$$I_d = I_o \left(e^{\frac{qV_d}{aKT}} - 1 \right), \tag{3.1}$$

where:

I_d = the current passing through the diode,
I_o = the reverse or dark saturation current of the diode at temperature T,
T = the absolute temperature in K,
a = the diode ideality factor which varies between 1 and 2,
q = the elemental charge of one electron measured as 1.6×10^{-19} C, and
V_d = the voltage across the diode.

3.2   Dark and Illumination Changes

As mentioned above, the variation of the light level gives rise to the changes in
the electrical characteristics of the diode. This change can be identified by the I-V
characteristics curve in Fig. 3.2 under illumination and in the dark. When the p-
n junction is illuminated, the I-V curve shifts downward when the p-n junction is

Fig. 3.3 Equivalent circuit
of an illuminated solar cell

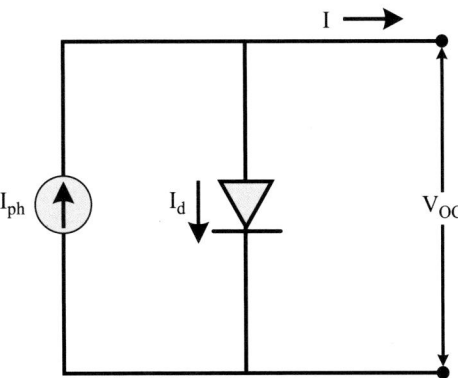

under illumination, and the generated or illuminated current is proportional to the intensity of the sunlight.

3.3   Ideal Equivalent Circuit of a Solar Cell

The ideal or perfect equivalent circuit (Fig. 3.3) of a solar cell contains a current source representing the illumination current, and a diode in parallel representing the dark current, where the load current is given using Eq. 3.2.

$$I = I_{ph} - I_d = I_{ph} - I_o \left(e^{\frac{qV_d}{aKT}} - 1 \right), \tag{3.2}$$

where,

I = the load current,
I_{ph} = the photon current generated by the solar cell, and
I_d = the current passing through the diode.

3.4   The I-V Curve

The current–voltage characteristic curve, also known as the I-V curve, is an essential characteristic of solar cells, which is used to illustrate the relationship between the voltage and the current produced by the solar module under the standard test conditions that have already been mentioned in Chap. 2. Under these conditions, the solar module considers a voltage-controlled current source, which means that the I-V curve will have a constant current value while the voltage value might change based on the connected load. As illustrated in Fig. 3.4, different resistive loads are connected to the same solar module. The point of intersection between the I-V curve

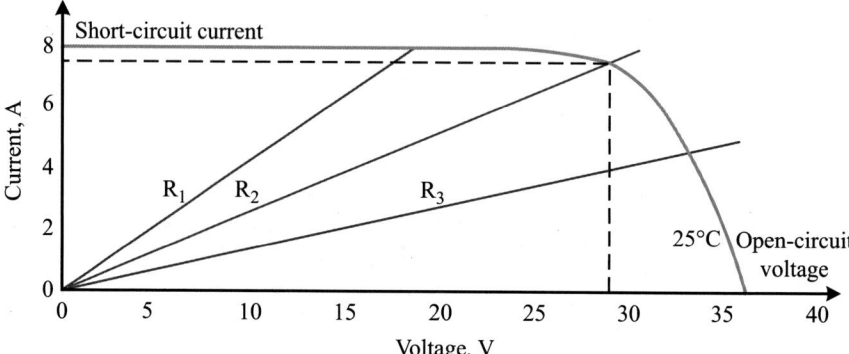

Fig. 3.4 The effect of resistance on the I-V characteristic curve of the solar module with different resistive values, where $R_3 > R_2 > R_1$

and the linear resistive load determines the operating power point. The electrical behavior of the solar module is plotted in this curve, where the x-axis represents the voltage and the y-axis represents the current. Three important points are to be noted from this graph, which are as follows:

1. Open-circuit voltage
2. Short-circuit current
3. Maximum power point

This curve can be plotted by connecting the module to a variable resistor (potentiometer) and by raising its resistivity from 0 ohms (short-circuit, where the curve has its Y-axis intersection) to a very large value (open circuit, where the curve has its x-axis intersection). At this range, the resistor passes a resistance value ($R2$ in Fig. 3.4) where the product of the current passing and the voltage drop reaches a maximum value. This point is called the maximum power point (MPP), and this is the point where the maximum power is obtained from the cell. The I-V characteristic curve found in terms of diodes is not linear, in contrast to resistors, as shown in Fig. 3.5.

When the cathode (n-type) is connected to the negative terminal of the voltage applied, and anode (p-type) is connected to the positive terminal, the current passes through the diode, which is considered a forward-biased state of the diode. But current does not flow as soon as the voltage is applied as there is a potential barrier between anode and cathode, which needs to be conquered. This minimum required voltage varies for different materials, for instance, 0.7 V for Silicon. Once the applied voltage exceeds the barrier voltage of the diode, an astonishing rise of current is evident for a small increase in the applied voltage. In the case of a reverse-biased situation, the cathode is connected to the positive terminal whereas the anode is connected to the negative terminal, which blocks the current flow through the diode. A small amount of current, known as leakage current, flows

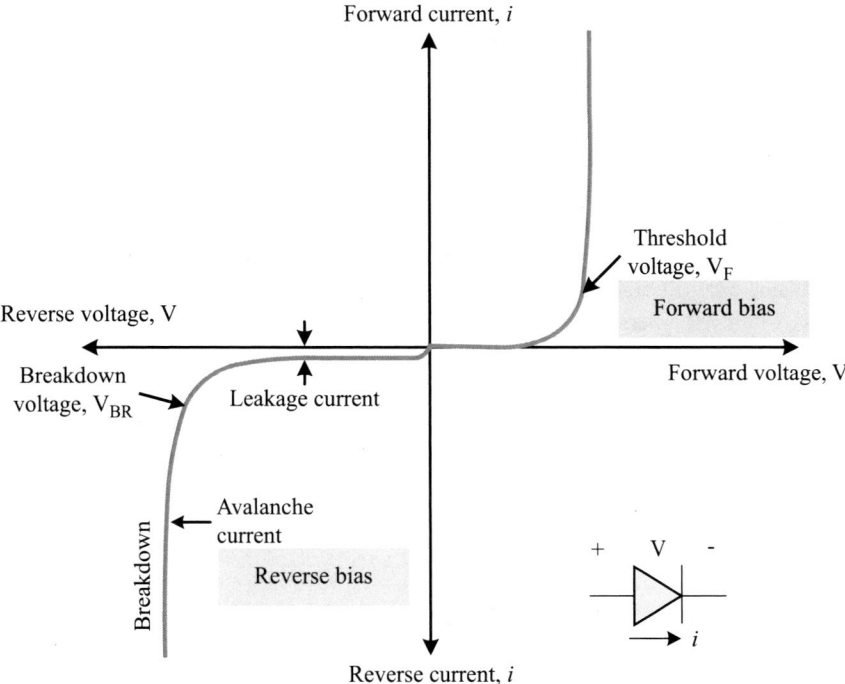

Fig. 3.5 I-V characteristic curve of a diode

across the circuit as no circuit is found in an ideal state as long as the voltage does not reach a particular voltage, named the breakdown voltage. The breakdown voltage is the largest reverse voltage which keeps the leakage current to a limited value. An exponential current flow is evident once the diode crosses the breakdown voltage as the voltage goes out of control at this stage. This large value of current is known as the avalanche current.

3.4.1   Open-Circuit Voltage

The open-circuit voltage, V_{oc}, as marked in Fig. 3.4, is the measured value of the voltage between the solar module's terminals when an infinite-resistance load is applied. This voltage is the highest possible voltage that the module can produce. The V_{oc} can be measured using a digital multimeter connected in parallel with the solar module terminals, showing a positive value as the output.

The output voltage value can be found in each module datasheet or determined from the I-V curve intersection between the x-axis and the curve. In Fig. 3.4, the open-circuit voltage value is found to be 36 V.

3.4.2   Short-Circuit Current

The short-circuit current I_{sc}, as shown in Fig. 3.4 is the maximum current that it can produce and is measured when the two terminals are connected without any load. This value can be obtained from the datasheet of each solar module or from the I-V curve, where the intersection between the y-axis and the curve represents the value of I_{sc}.

It is worth noticing that the module terminals' cables are designed to handle the high value of I_{sc}. Thus, the short-circuit test can be done by disconnecting the solar modules and testing it as a standalone system after ensuring that the multimeter range is equal or greater than the expected value of the short-circuit current. The measured value of the I_{sc} is very high, which is not recommended to be applied by any energy sources, unlike PV.

3.4.3   Maximum Power Point

As known in electrical engineering, the power extracted from DC energy sources is obtained from the product of the current and voltage ($P = V \times I$). Therefore, there will be no useful power supplied to the load when the circuit is operating in the short-circuit or open-circuit conditions because one of the factors will be equal to zero.

In PV technology, the load regulates the operating point, where the solar module delivers both current and voltage in terms of the resistive load connected to its terminals. Therefore, when a variable resistor (potentiometer) is applied to the solar module with an increasing value, the solar module keeps increasing the delivered power until it reaches the peak value and then starts declining. The critical point at which the module delivers the highest possible power is called the maximum power point (MPP), and its x-axis and y-axis coordinates are known as voltage at MPP, V_{mp} and current at MPP, I_{mp}, respectively. In the example given in Fig. 3.4, R_2 is the resistive load value at which the module can operate at its MPP, where the values of I_{mp} and V_{mp} are 7.5 A and 28 V, respectively.

The I-V characteristics curve with open-circuit voltage, short-circuit current, and maximum power point is illustrated in Fig. 3.6.

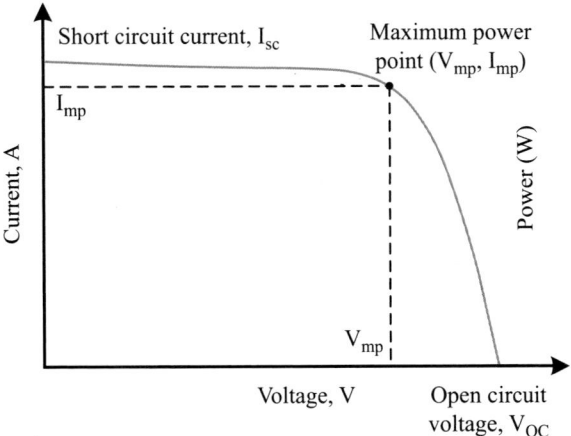

Fig. 3.6 I-V characteristic curve of a solar module highlighting short-circuit current, open-circuit voltage, and maximum power point

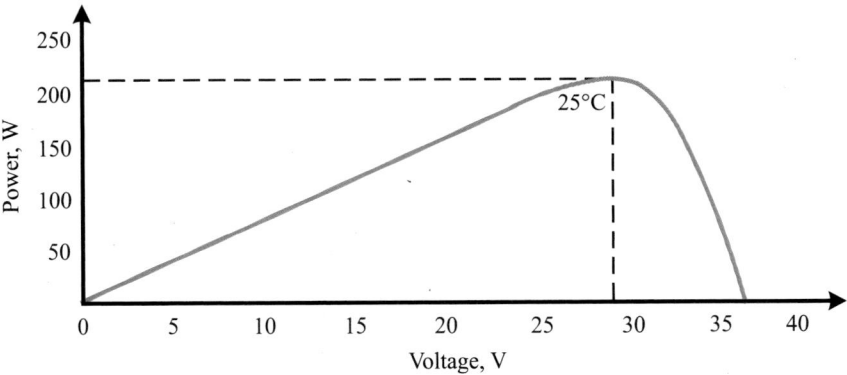

Fig. 3.7 P-V curve of the solar module

Table 3.1 Module specification

$V_{oc}(V)$	$I_{sc}(A)$	$I_{mp}(A)$	$V_{mp}(V)$	$M_{pp}(W)$
36	7.9	7.5	28	210

3.5 The P-V Curve

The power–voltage or P-V curve (Fig. 3.7) is provided by the manufacturer in the module datasheet, where the current is multiplied by the voltage to give the expected output power at every voltage dropped on the load. From both curves, we can extract several important values and list them in Table 3.1 as the following:

3.6 Effect of Temperature

As discussed in Chap. 2, the solar module temperature is much warmer than the ambient temperature. All specifications and curves are taken under a cell temperature of 25 °C as one part of the standard test conditions (STC). However, the solar module's operating conditions do not necessarily match the STC, especially in its temperature, where the modules can be installed in extreme cold or hot weather, such as in Canada or Mali, respectively. Therefore, it is important to define a few factors to define the relationship between the temperature and characteristics of the solar module, such as current, voltage, and power. The voltage, current, and power temperature coefficients discussed in the next three sections denote the relationship of these parameters with the temperature.

3.6.1 Voltage Temperature Coefficient (%/°C)

The average value of the voltage temperature coefficient varies from -0.27 to $-0.35\%/°C$. The negative sign reflects the negative impact on the open-circuit voltage as the temperature rises. This means that for 1 °C of temperature rise, the open-circuit voltage will decrease by a certain percentage in the mentioned range. This factor is crucial due to the nature of the power electronic devices that are always connected to PV modules, such as inverters. Each inverter has a threshold voltage value under which the inverter cannot operate. Thus, the solar array must be well-designed to compensate for voltage drops resulting from high temperatures to keep the power electronic devices in their normal operation ranges. For the purpose, the lowest possible voltage of power electronics devices or the highest possible temperature where the voltage is at its minimum value is considered.

The percentage of gain or loss of the voltage can be calculated using Eqs. 3.3 and 3.4:

$$Operating\ voltage\ in\ \% = V_{T-coefficient} \times (T_{cell} - 25) + 100. \qquad (3.3)$$

Therefore, the actual operating voltage is:

$$V_{specific\ temperature} = Operating\ voltage\ in\ \% \times V_{STC}. \qquad (3.4)$$

3.6.2 Current Temperature Coefficient (%/°C)

The current temperature coefficient always has a positive value oscillating between 0.047 and 0.055%/°C in modern polycrystalline models. This means that for 1 °C of temperature rise, the short-circuit current will increase by a specific percentage in the mentioned range. However, the positive effect of the temperature rise on the current can be neglected compared to the negative voltage effect. The percentage of gain or loss of the current can be calculated using Eqs. 3.5 and 3.6.

$$Operating \ current \ in \ \% = I_{T-coefficient} \times (T_{cell} - 25) + 100. \quad (3.5)$$

Therefore, the actual operating current is:

$$I_{specific \ temperature} = Operating \ current \ in \ \% \times I_{STC}. \quad (3.6)$$

3.6.3 Power Temperature Coefficient (%/°C)

The impact of the power temperature coefficient is illustrated in Fig. 3.9. The output power is inversely proportional to the temperature. The coefficient value varies from -0.45 to $-0.3\%/°C$, where the coefficient value defines the quality of the silicon-based module. The negative sign implies that with 1 °C temperature rise, the output power of the solar cell decreases by a certain percentage within the mentioned range. The percentage of gain or loss of the power can be determined using Eq. 3.7.

$$Operating \ power \ in \ \% = P_{T-coefficient} \times (T_{cell} - 25) + 100. \quad (3.7)$$

In standard solar cell datasheets [2], it is possible to find the values of the voltage, current, and power coefficients in terms of %/K, where K is the temperature in Kelvin. However, the difference in each degree temperature is the same in Celsius scale and Kelvin scale. Therefore, the same value can be used irrespective of the values expressed in terms of %/K or %/°C.

Figures 3.8 and 3.9 illustrate the difference between the STC I-V and P-V characteristic curves that the manufacturers give in the datasheet and the actual

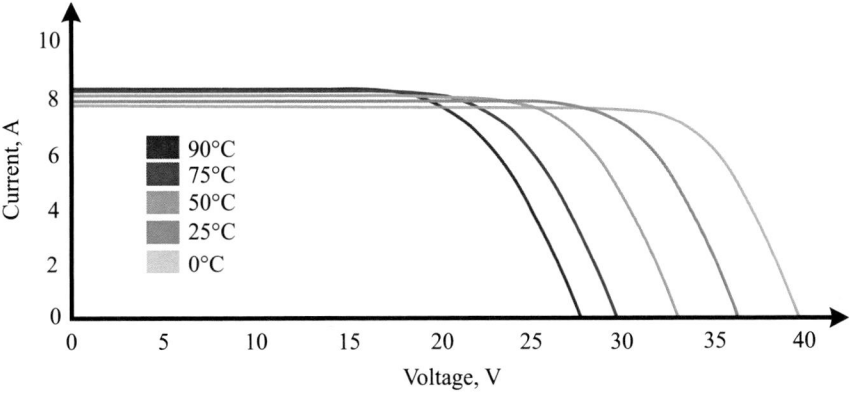

Fig. 3.8 I-V curve at different temperatures

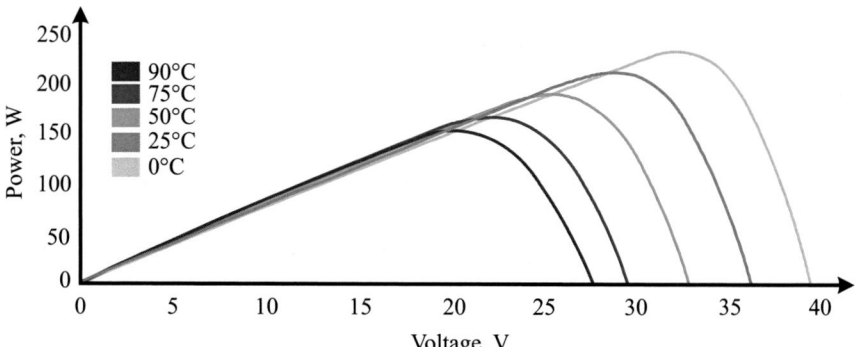

Fig. 3.9 P-V curve at different temperatures

Table 3.2 Actual gain and losses of current, voltage, and power based on V_{oc}, I_{sc}, and power coefficient at different temperatures

Ambient temperature, °C	Cell temperature, °C	V_{oc} coefficient, %/°C	Actual operating voltage percentage of the nominal voltage, %	I_{sc} coefficient, %/°C	Actual operating current percentage of the nominal current, %	Power coefficient, %/°C	Actual operating power percentage of the nominal power, %
−20	6	−0.29	105.51	0.049	99.069	−0.39	107.41
−10	16		102.61		99.559		103.51
0	26		99.71		100.049		99.61
10	36		96.81		100.539		95.71
20	46		93.91		101.029		91.81
30	56		91.01		101.519		87.91

curves that the module works at; based on the operating cell temperature, the irradiance is fixed to 1000 W/m^2.

In some rare situations, it is possible that the designer has to design his system without sufficient information. Therefore, using the table created by the American National Electric Code (NEC 2020) provides the designer with a voltage correction factor for multicrystalline and monocrystalline modules based on a wide range of ambient temperatures [3]. Table 3.2 shows the response of voltage, current, and power of a solar module at different temperatures, while it can be visualized by Fig. 3.10.

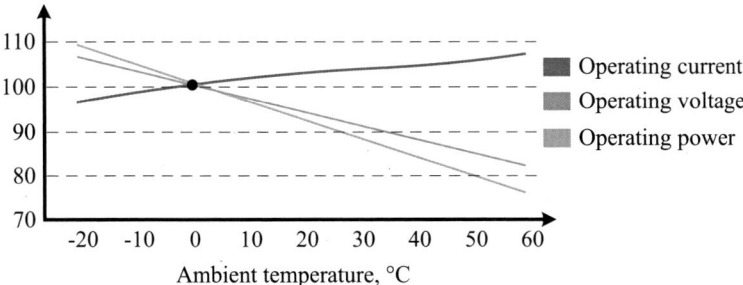

Fig. 3.10 The percent gains/losses of the operating current, operating voltage, and operating power of a solar module at different temperatures

3.7 Effect of Irradiance

Solar irradiance factor (x) has a simpler and more straightforward effect on solar cells or modules, where it has a negligible effect on V_{oc} and a proportional effect on I_{sc} as illustrated in Eq. 3.8 and Fig. 3.11.

$$I_{sc_x \ kW/m^2} \equiv x \times I_{sc_1 \ kW/m^2}, \tag{3.8}$$

where x varies from 0 to 1.25. The current in this equation refers to the short-circuit current at a specific irradiance measured in kW/m^2.

3.8 Fill Factor

The fill factor (FF) can be used to compare several modules together where the higher the FF, the better is the electrical performance by the PV module. FF is the ratio of the maximum power produced by the solar module to the product of V_{oc} and I_{sc} as Eq. 3.9 shows.

$$FF = \frac{P_{max}}{I_{sc} \times V_{oc}} = \frac{I_{mpp} \times V_{mpp}}{I_{sc} \times V_{oc}}. \tag{3.9}$$

The ratio between the surface area covered by the filled rectangular and the outer dashed rectangular of the I-V curve also represents the fill factor. Figure 3.12 illustrates the fill factor of a solar module.

The value of the FF ranges from 50–82%. For a Silicon PV cell, the FF is usually about 80%. The value increases with a higher shunt resistance and a lower series resistance. A higher value of FF is desired because it is a direct indicator of the quality of the PV cell.

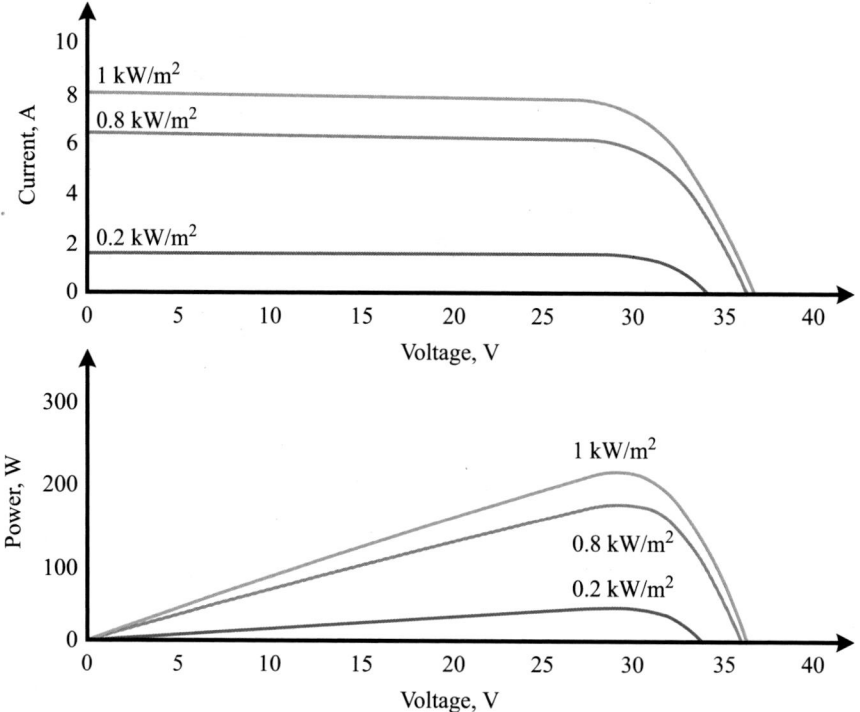

Fig. 3.11 I-V and P-V curves under different irradiance values

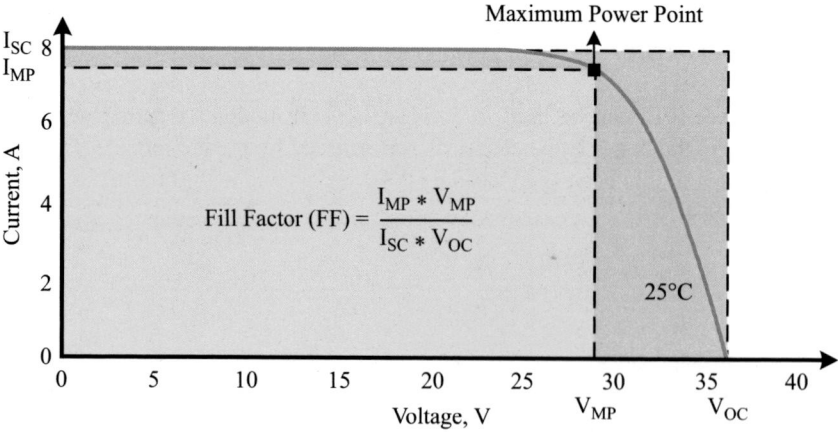

Fig. 3.12 Representation of fill factor of a solar module

3.9   Resistance

The ideal equivalent solar cell discussed previously does not take into consideration any of the actual losses caused by the cell defects or other ohmic losses, which can be represented by two resistors, one connected in parallel with the diode or current source and the other is in series as illustrated in Fig. 3.13.

3.9.1   Series Resistor

Since the actual solar cell contains different factors that add losses to both voltage and current, it is essential to address these losses and include them in the ideal cell equivalent circuit. Series resistor represents the losses caused by the ohmic resistivity of the metallic grid placed on the upper side of the cell and the busbars that collect the current produced by the cell. The supplied voltage is divided among the series resistors, causing power dissipation in the form of heat, also known as Ohm's effect, in each resistor. The total current passing through a circuit, including series resistors, is illustrated in Eq. 3.10 [4]:

$$I = I_{ph} - I_s \left[\left(e^{\frac{q(V + IR_s)}{aKT}} \right) - 1 \right] + \frac{V + IR_s}{R_{sh}}. \tag{3.10}$$

where,

I = the output current,
I_{ph} = the value of the current source,
I_s = the reverse saturation current,
q = the value of an unit electric charge,
V = the terminal voltage of the circuit,
a = the ideality factor,
K = the Boltzmann constant,
T = the operating temperature, and

Fig. 3.13 Actual solar cell equivalent circuit

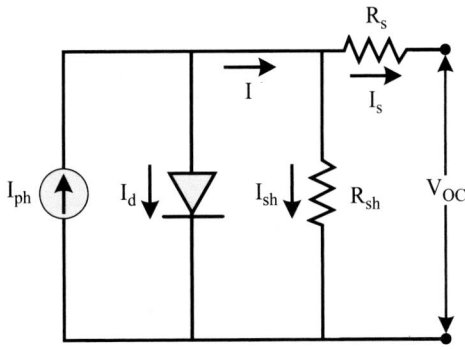

R_{sh} = the shunt resistance in the equivalent circuit.

The recommended value for the equivalent series resistance, R_s, is zero, where all the metallic contacts are ideal conductors; therefore, no voltage drop occurs, and all the voltage provided by the solar cell/module can be transferred to the load. It is worth noting that both R_s and R_{sh} have a massive effect on the FF.

3.9.2 Parallel Resistor

The value of the parallel resistor in the equivalent circuit indicates the leaking currents along the solar cell edges. Ideally, parallel resistors are supposed to be equal to infinity, and that is why they do not show up in the ideal equivalent circuit. However, due to the solar cell defects in the manufacturing process, the solar cell's actual equivalent circuit has relatively large resistances that sink small values of currents. This loss of current gives rise to the drop of maximum power point value. The parallel resistor, also known as a shunt resistor, has a minimal effect on the V, unlike its major effect on the I. The value of I can significantly decrease if the value of the shunt resistor is low, as described in Eq. 3.11:

$$ I = I_{ph} - I_s(e^{\frac{qV}{aKT}} - 1) + \frac{V}{R_{sh}}. \qquad (3.11) $$

3.10 Conclusion

This chapter provides an extensive explanation of the properties of solar PV cells. The relationship between current and voltage for diodes has been shown using the I-V curve. The shape of the I-V curve changes at different levels of light. This diode property has been explained with necessary graphs and figures, which compare the behaviors of diodes at different levels of light. This chapter has also emphasized three important parameters of the PV cells, i.e., open-circuit voltage, short-circuit current, and maximum power point. The P-V curve, depicting the relationship between the power and voltage of the PV cell is also elaborately explained. The properties of solar cells vary with temperature as they are temperature-dependent. The change of voltage, current, and power with the temperature has been discussed in terms of coefficients, by which these operating values can easily be found. Necessary graphs and data tables have been provided to give a precise idea to the readers about the variation of properties of solar cells. The effect of irradiance on the solar cells along with fill factor has been discussed in this chapter, followed by variations caused by different arrangements of resistors. In conclusion, this chapter provides the readers with a clear understanding of solar cells' properties and internal designs.

3.11 Exercise

1. What is the difference between a diode and a solar cell?
2. Draw the I-V characteristic curve of a diode with proper explanation.
3. What information can you obtain from the I-V and P-V curves of a solar cell?
4. Define the following with their units, if applicable:
   (a) Open-circuit voltage
   (b) Short-circuit current
   (c) Maximum power point
   (d) Current at maximum power
   (e) Voltage at maximum power
   (f) Fill factor
   (g) Voltage temperature coefficient
   (h) Current temperature coefficient
   (i) Power temperature coefficient
5. Draw the equivalent circuit of a solar cell and define the components.
6. How does the solar irradiance affect the open-circuit voltage and short-circuit current of a solar cell?
7. Why do the voltage, current, and power of a solar cell vary based on the temperature?
8. Observe the following data for a solar PV cell:

   • Open-circuit voltage = 0.63 V
   • Short-circuit current = 8.94 A
   • Voltage at maximum power = 0.54 V
   • Current at maximum power = 8.37 A

   Determine the fill factor of the solar cell. [Answer: 0.802]
9. What are the factors that influence the output of the solar cell? Briefly explain their impacts.

References

1. B. Smith, How differing levels of light affect different types of solar panels
2. Q6LMXP3-G3. https://www.q-cells.eu/uploads/tx_abdownloads/files/Hanwha_Q_CELLS_Data_sheet_Q6LMXP3-G3_2013-04_Rev01_EN_01.pdf
3. National Fire Protection Association (NFPA), *NFPA 70—National Electric Code* (2020)
4. H. Koffi, A. Yankson, A. Hughes, K. Ampomah-Benefo, J. Amuzu, Determination of the series resistance of a solar cell through its maximum power point. Afr. J. Sci. Technol. Innovation Dev. **12**(1–4), 03 (2020)

Solar Resources

4

4.1   Introduction

The Sun is the ultimate energy source for the Earth and the basis of the solar energy technologies used to generate clean energy. In order to study solar PV technology, it is imperative to gain a solid understanding of the Sun, the relation between the Sun and the Earth, and the solar resources available to us for use.

4.2   The Sun and Its Radiation

Being the only star in our solar system, the Sun is proudly the only source of energy for all the planets revolving around it. The Sun is a burning ball of gases, primarily Hydrogen and Helium, but also contains a mixture of Oxygen, Carbon, Nitrogen, Iron, etc., totaling about 67 elements. Inside the core of the Sun, the gravitational forces of the Sun generate a high temperature and pressure. At its core, the temperature of the Sun is almost 27 million °F (15 million °C). The Hydrogen atoms are fused together to produce Helium atoms in the process of nuclear fusion. Nuclear fusion reaction continuously occurs within the Sun, which liberates a colossal amount of energy radiating outward from the star. Due to the intense heat, the gaseous atoms get charges into ions and thus form a plasma. The overall radiation of the Sun mostly contains magnetic, heat, and light energies.

4.3   Radiation Spectrum

Solar radiation is composed of heat and light energies and ultraviolet (UV) radiations. The solar spectrum constitutes 6.6% of UV rays (less than 380 nm), 44.7% of visible light (380–780 nm), and 48.7% of infrared (IR) rays (more than 780 nm, which is technically heat energy). Figure 4.1 demonstrates the solar radiation

© The Author(s), under exclusive license to Springer Nature Switzerland AG 2022
Y. Abou Jieb, E. Hossain, *Photovoltaic Systems*,
https://doi.org/10.1007/978-3-030-89780-2_4

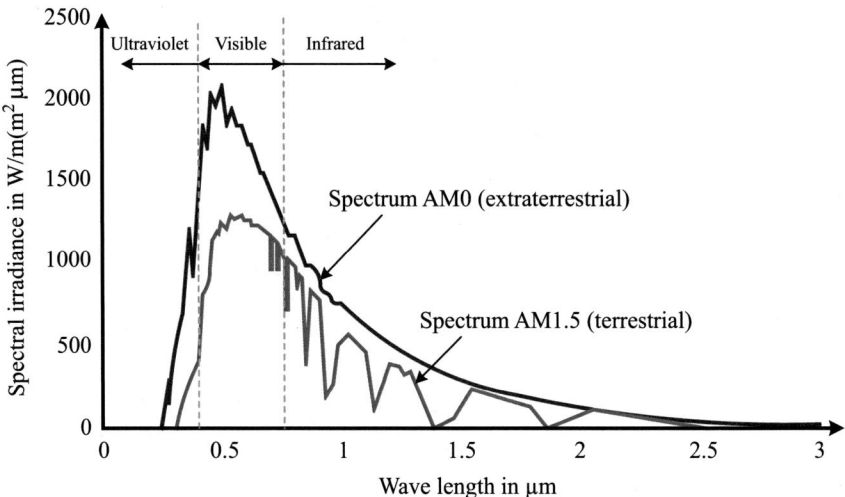

Fig. 4.1 The solar radiation spectrum

spectrum, which corresponds to the spectral irradiance for each wavelength of the solar spectrum. The spectrum is considered with respect to air mass (AM) index, a term that will be elaborated on later in this chapter. The red curve shows the spectrum AM0, which is the extraterrestrial radiation that is received at the outer periphery of the Earth's atmosphere from the Sun. The blue line represents the spectrum AM1.5, which is the terrestrial radiation received on the ground. AM1.5 is lower than AM0 because of the atmospheric losses of solar radiation.

4.4   Global Insolation

It has been stated earlier that the equatorial region of the globe receives the maximum amount of sunlight during the daytime. The places closer to the two poles are relatively deprived of enough solar irradiation. Solar irradiance can be defined as the power of solar radiation befalling per unit surface area. It is measured in W/m^2. On the other hand, solar irradiation is defined as the energy of solar radiation over a definite period. It is measured in kWh/m^2. Insolation is the amount of solar radiation energy incident on a specific surface area in a given period. It can be denoted as the average irradiance in kWh per unit area per day ($kWh/m^2/day$).

4.5   Direct and Diffused Radiation

The sunlight that reaches the surface of the Earth unobstructed without being reflected, deviated, or absorbed in any of the atmospheric layers or other terrestrial bodies is known as direct radiation. Such beams are ideally suited for concentration for harvesting solar energy.

On the contrary, the sunlight that is bounced back from any atmospheric or terrestrial obstacle is known as diffused radiation. Such indirect beams cannot cast sharp shadows and are unsuitable for concentration [1].

The sum of the direct and diffused radiations gives the total solar radiation on a horizontal surface.

4.6   Solar Resources Definitions

The *global horizontal irradiance (GHI)* refers to the overall amount of cumulative solar rays that fall on a surface that is horizontal to the Earth. It is made up of the *direct normal irradiance (DNI)*, the *diffused horizontal irradiance (DHI)*, and the *ground-reflected radiation*. Since most solar modules are mounted at a tilt angle, GHI is used as a starting point to calculate the solar irradiance on a tilted surface (solar array) to determine the best tilt angle that achieves the high DC electrical energy yield.

The DNI is a crucial aspect of concentrating solar PV technology, which refers to how much direct solar radiation a surface receives per unit area that is always oriented directly toward the sunbeam. The relationship between GHI, DHI, and DNI can be represented in terms of the zenith angle z as Eq. 4.1. The concept of zenith angle will be explored later in this chapter.

$$GHI = DHI + DNI \times \cos z. \tag{4.1}$$

The ground-reflected radiation is the total solar radiation that is reflected from the surface of the Earth, and it is measured using a ratio called *albedo*. The albedo, also referred to as the *whiteness of a surface*, is a ratio of the amount of solar irradiation reflected from a surface to the total amount of irradiation befalling it. The albedo ranges from 0 to 1 and varies with the location, season, and time of day. Completely white objects have an albedo of 1, as they can reflect 100% of the radiation incident on them, whereas completely black objects have an albedo of 0, as they can absorb 100% of the radiation incident on them.

The *global tilted irradiance (GTI)* refers to the total amount of irradiance befalling a surface with a definite tilt and azimuth angle. The GTI is the sum of the scattered, direct, and reflected radiation. The GHI, DNI, DHI, ground-reflected radiation, and GTI are measured in kWh/m^2 per period.

Figure 4.2 illustrates how the Sun rays are incident on the Earth in various ways—some rays are directly radiated onto the solar module, some rays are reflected from the ground, and some are diffused through the clouds. A significant portion of Sun rays is reflected away from the clouds and also absorbed by the clouds.

The PV potential of the world can be observed from Fig. 4.3, which shows that Africa has the highest observed PV power potential. Figure 4.4 maps out the variation of the DNI across the globe, while Fig. 4.5 illustrates the variation of the

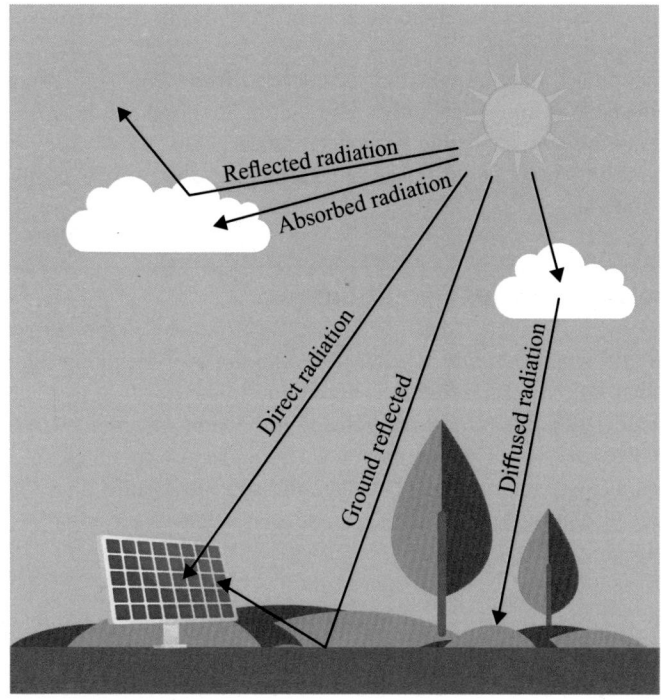

Fig. 4.2 The many ways in which the rays of the Sun reach the Earth or are deflected away

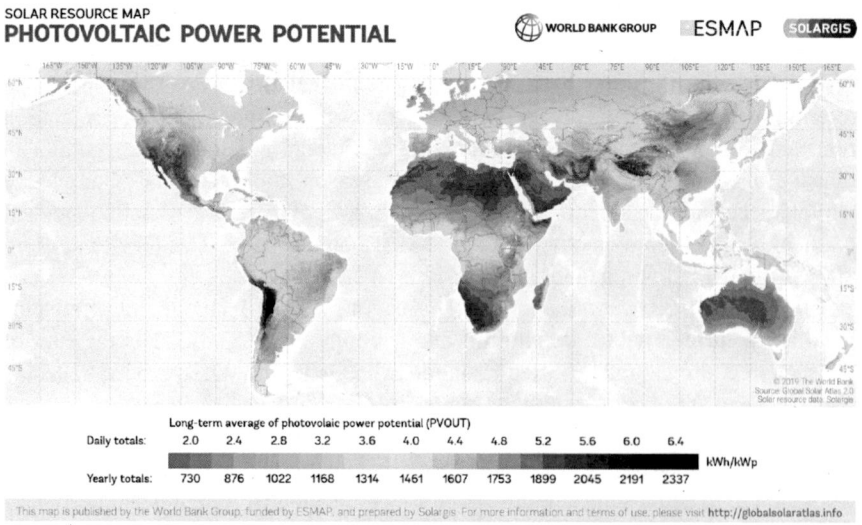

Fig. 4.3 Global Photovoltaic Power Potential. Map obtained from the "Global Solar Atlas 2.0," a free, Web-based application developed and operated by the company Solargis s.r.o. on behalf of the World Bank Group, utilizing Solargis data, with funding provided by the Energy Sector Management Assistance Program (ESMAP). For additional information: https://globalsolaratlas. info

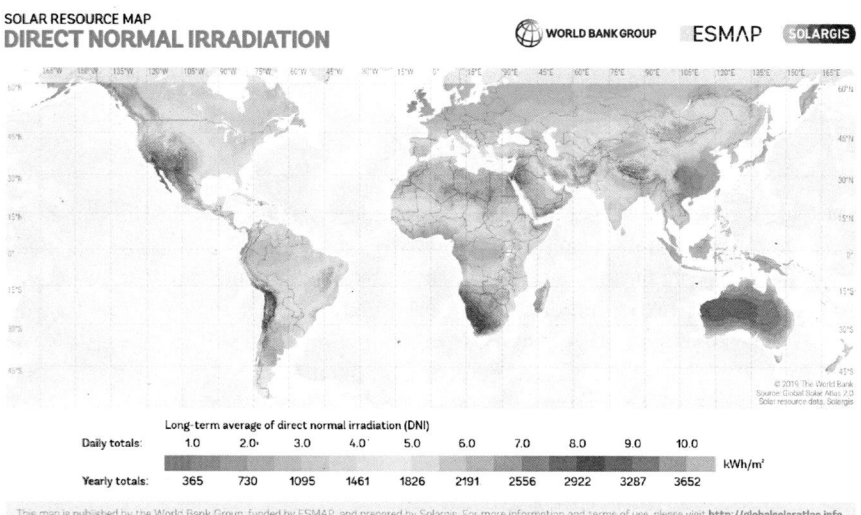

Fig. 4.4 Global Direct Normal Irradiation (DNI). Map obtained from the "Global Solar Atlas 2.0," a free, Web-based application developed and operated by the company Solargis s.r.o. on behalf of the World Bank Group, utilizing Solargis data, with funding provided by the Energy Sector Management Assistance Program (ESMAP). For additional information: https://globalsolaratlas. info

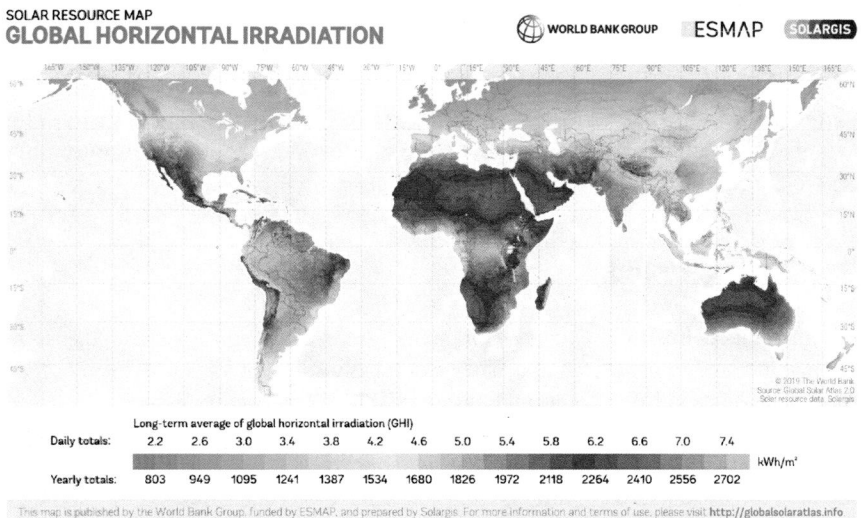

Fig. 4.5 Global Horizontal Irradiation (GHI). Map obtained from the "Global Solar Atlas 2.0," a free, Web-based application developed and operated by the company Solargis s.r.o. on behalf of the World Bank Group, utilizing Solargis data, with funding provided by the Energy Sector Management Assistance Program (ESMAP). For additional information: https://globalsolaratlas. info

GHI around the globe. It can be observed that the DNI is the highest around the two tropics, while the GHI is the highest on either side of the equatorial belt.

4.7   Atmospheric Effects

The output power of solar modules is largely dependent upon climatic and atmospheric vagaries. Rainfall, snowfall, storms, wind flow, dust accumulation, and other environmental phenomena cause variations in the module output power. A study in Alberta, Canada, noted that due to the accumulation of snow on the modules, the energy loss per annum ranges from 1.6% at an optimal tilt (53°) to 5.3% at a lower tilt (15°). The deposition of dust and debris on the modules also lowers the output by blocking out solar radiation.

4.8   Earth's Radiative Equilibrium

Radiative equilibrium refers to a condition in which the total thermal radiation exiting from and entering a body is equal. The radiative equilibrium of the Earth signifies the concept that the amount of solar energy entering the Earth is equal to the energy leaving from the Earth. This is also known as the Earth's energy budget. Of all the incoming solar radiation, about 19% is absorbed by the clouds, water vapor, dust, and other components in the atmospheric layers, 51% is absorbed by the land and oceans of the Earth, and nearly 30% is reflected back into space. The scattering phenomenon also sends some rays back into space and some to the Earth. But the scattered rays are not strong enough to produce a significant output in PV modules. Figure 4.6 depicts the energy budget of the Earth, which is a technical term for the amount of incoming, absorbed, reflected, scattered, and outgoing energy from the Sun to the Earth.

4.9   Air Mass Index

Air mass is defined as a large volume of atmospheric air that has the same temperature and moisture content. They extend thousands of kilometers across their source regions and from the ground level to about 16 km high up in the atmosphere.

The air mass index or air mass coefficient is the direct path through the atmosphere that a light ray traverses. It is generally used to assess the performance of PV cells under standard conditions. It is denoted by AM and followed by a number. The process of calculating the air mass index is shown in Fig. 4.7. For example, the most common denotation is AM1.5 to characterize terrestrial PV modules. The air mass index is calculated using Eq. 4.2.

$$AM = \frac{L}{L_o},\tag{4.2}$$

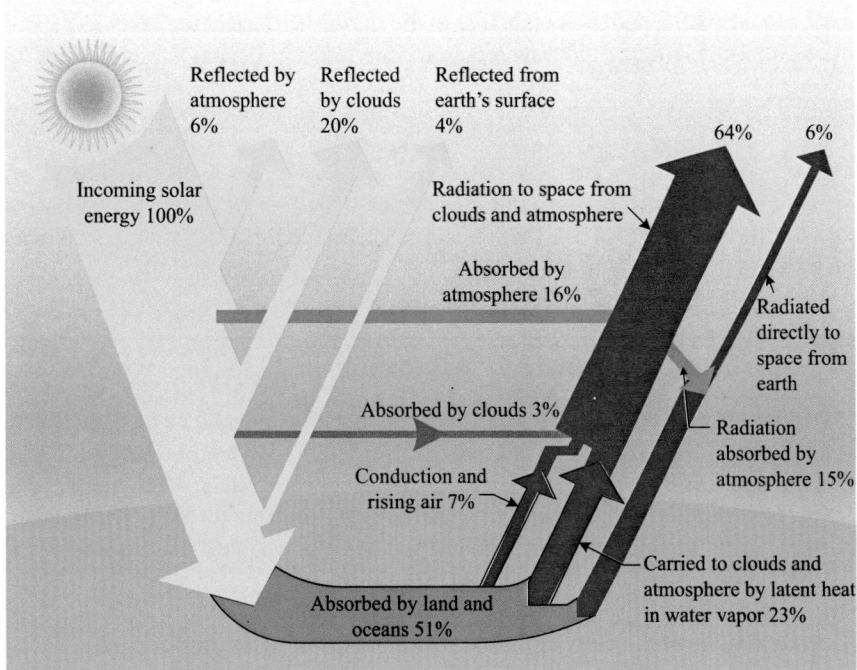

Fig. 4.6   The energy budget of the Earth [2]

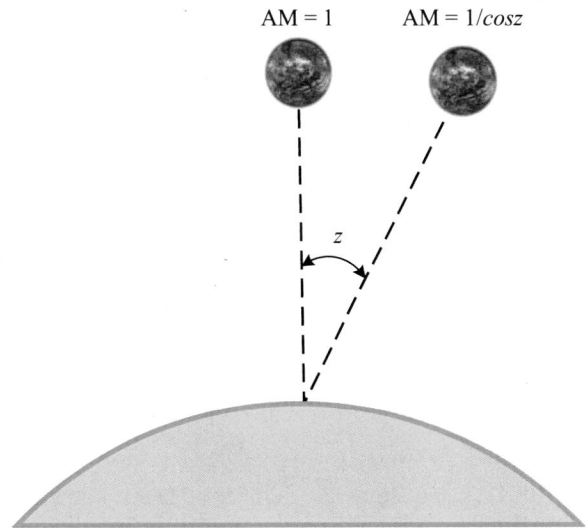

Fig. 4.7   The process of calculating the air mass index

where L is the path length through the atmosphere at any angle z with the normal, and L_o is the path length perpendicular to the Earth's surface at sea level.

The air mass index can also be represented in terms of the angle z as Eq. 4.3.

$$AM = \frac{1}{\cos z}, \tag{4.3}$$

where z connotes the zenith angle in degrees. However, this equation does not consider the curvature of the Earth's surface. The modified equation considering the Earth's curvature is Eq. 4.4 [3,4].

$$AM = \frac{1}{\cos z + 0.50572(96.07995 - z)^{-1.6364}}. \tag{4.4}$$

AM0 is the extraterrestrial air mass, i.e., the radiation incident on the atmosphere of the Earth. AM1 indicates the air mass when the Sun is directly overhead, i.e., at the solar noon. Si PV cells have been found to be more efficient at AM1 compared to AM0. However, the output power is higher at AM0 compared to AM1. The standard air mass on the ground level is referred to as AM1.5G, where G stands for global.

Standard Air Mass
A standard value of the air mass index is chosen for testing the performance of solar PV systems. This standard is the AM1.5, which gives the amount of total radiation incident on the Earth's surface as $1 \, \text{kW/m}^2$. For AM1.5, the solar zenith angle is 48.2°, i.e., the angle measured clockwise from the normal on the surface of the Earth.

4.10 Solar Constant

The amount of solar energy received per unit area on a surface normal to the Sun's rays at a distance of one astronomical unit (au) from the Sun is known as the solar constant. The value of the solar constant is $1.366 \, \text{kW/m}^2 \approx 1.37 \, \text{kW/m}^2$. The solar constant is typically calculated at the top of the atmosphere without considering the atmospheric losses of the sunlight. The knowledge of the solar constant helps to anticipate the input power on a surface on the Earth and is particularly important in calculating the efficiency of a solar module.

4.11 Peak Sun Hour

Peak Sun hours (PSH) is defined as the number of hours when the average GHI value will be $1000 \, \text{W/m}^2$ in a day. The amount of solar irradiance is maximum at noon, which can be more than $1000 \, \text{W/m}^2$, whereas, in other times, this value can be more or less than $1000 \, \text{W/m}^2$ (although it is less probable to have an irradiance

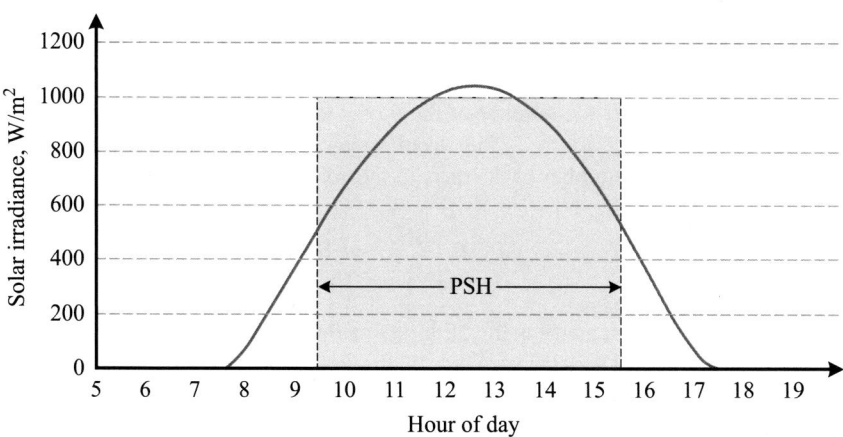

Fig. 4.8 Variation of the solar irradiance throughout the day and the concept of peak Sun hour

exceeding $1000\,\text{W/m}^2$). Figure 4.8 illustrates the PSH at any location based on the solar irradiance of any day. For the given Fig. 4.8, the value of PSH is $6\,\text{kWh/m}^2$ per day, which means for 6 h that day, the average value of solar irradiance is $1\,\text{kW/m}^2$ or $1000\,\text{W/m}^2$. This duration is from 09:30 to 15:30. In this particular example, albeit the PSH varies from place to place. Even in one particular place, the PSH varies every day, and usually, the monthly average is calculated for getting an idea about the PSH of any place, as shown in Eq. 4.5. The unit of PSH is kWh/m^2 per day.

$$PSH = \frac{Avg.\ GHI\ of\ a\ month}{No.\ of\ days\ in\ that\ month}. \qquad (4.5)$$

4.12 Sun Angles

Since the Earth is a spherical object, the coordinates of the Sun from the Earth as a reference are best described in terms of the spherical coordinate system. The fixed perpendicular line from the center of the spherical coordinate system is known as the zenith. In the paragraphs that follow, we are going to get introduced to the most pivotal concepts pertinent to solar geometry, which will help us to further understand the accurate positioning of solar PV modules. The knowledge of these angles and terminologies is important for the selection of the appropriate location and direction for installing solar PV systems to ensure maximum power output from the system.

Declination Angle, δ

The angle between the direction of the Sun ray and the equatorial plane of the Earth is known as the Sun declination angle, δ. Conventionally, the declination angle is measured in degrees. If the number of the day is known in the solar year, then the declination angle of the Earth on that specific day can be calculated using Eq. 4.6, where N refers to the number of the days in the year.

$$\delta = 23.45 \sin \left[\frac{360}{365} (284 + N) \right]. \tag{4.6}$$

For example, 29th January is the 29th day of the year. So, the declination angle on 29th January is $-18.298°$. For each of the 365 days in a year, if the δ values are calculated and plotted in a graph, Fig. 4.9 will be obtained. The figure shows that the declination angle follows the shape of a bell curve, with the highest value at $+23.45°$ on 21st June, and the lowest value at $-23.45°$ on 21st December. The positive angles indicate that the northern hemisphere of the Earth is tilted toward the Sun, which is the case from March to September. On the contrary, the negative angles imply that the southern hemisphere of the Earth is tilted toward the Sun, from September to March.

Zenith Angle, z

The position of the Sun measured vertically downward from the zenith is known as the zenith angle z, which corresponds to the polar angle θ in the spherical coordinate system. In other words, this is the angle between a solar ray and the perpendicular from the horizontal plane.

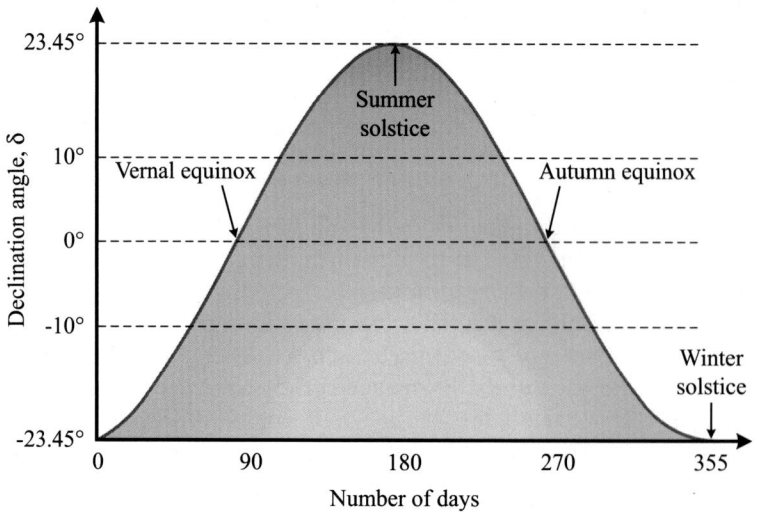

Fig. 4.9 Variation of the declination angle throughout the year

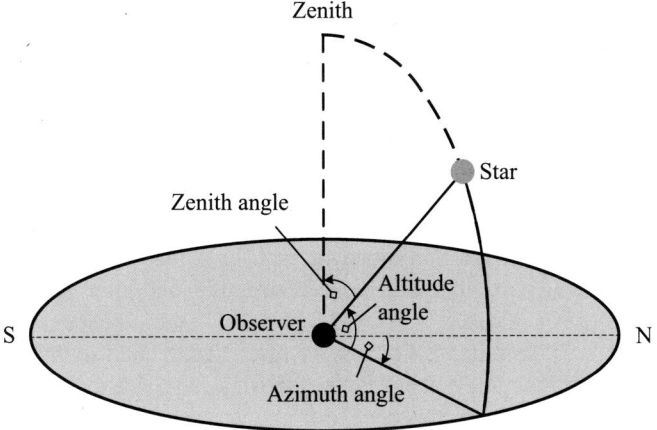

Fig. 4.10 The solar angles

Altitude Angle, α

The angular position of the Sun measured vertically upward from the horizon is known as the solar elevation angle or the altitude angle, α, of the Sun. In other words, this is the angle between the Sun ray and its projection on a horizontal plane on the Earth. It is also known as the inclination angle. From Fig. 4.10, it is evident that the sum of the altitude angle and the zenith angle is 90°, as shown in Eq. 4.7. So, they are complementary angles to each other.

$$z + \alpha = 90°. \tag{4.7}$$

The mathematical expression to determine the altitude angle of the Sun at any location from the Earth is provided in Eq. 4.8.

$$\sin \alpha = \cos z = \sin L \sin \delta + \cos L \cos \delta \cos h, \tag{4.8}$$

where L is the latitude at the specified location and h is the hour angle.

Azimuth Angle, θ

Imagine the Sun's perpendicular projection on the horizontal plane of the Earth. The horizontal angular distance between the projection and the true north direction is known as the solar azimuth angle. An azimuth angle of 0° refers to the true north direction, 90° refers to the east, 180° refers to the south, and 270° refers to the west.

Solar Noon

The solar noon refers to the time when the Sun is at the maximum altitude from the horizon of the Earth. This might instigate the idea that the solar noon is at 12 p.m.; however, the solar noon and the local noon are not the same. Not all locations on the Earth have their solar noon at exactly 12 p.m. due to the differences in the time zone and daylight saving hours. This implies that the solar noon is when the Sun is exactly located at the zenith of a certain location. For example, the solar noon at Klamath Falls, Oregon, USA, is recorded to be at 1:12 p.m. on August 9th, 2021.

Hour Angle, h

Imagine a typical wall clock, where the 12 h are spaced equally on a circle measuring 360°. The hour angle is simply the angle between the solar noon and the time denoting the position of the Sun along its path in the sky. The hour angle is 0° at 12 p.m. Before 12 p.m., the hour angle is negative, and after 12 p.m., the hour angle is positive. For instance, 9 am is 3 h prior to 12 p.m.; hence, the hour angle is $-3 \times 15°$ or $-45°$. Similarly, 4 p.m. is 4 h after 12 p.m., and so the hour angle is $4 \times 15°$ or 60°. The hour angle at any hour of the day can be easily determined using the aforesaid calculation, which is derived from the more formal formula in Eq. 4.9.

$$h = \frac{360}{24}(t - 12) = 15(t - 12), \tag{4.9}$$

where t is the current time in hours.

4.13 Sun Path

At a particular geographical location, the position of the Sun in the sky can be ascertained using a *solar chart*, also known as a *Sun path diagram*. The diagram provides the Sun's position in terms of the azimuth angle and the altitude on an hourly and monthly basis for any specific location on the Earth. Sun path diagrams are unique to each latitude of the Earth. This diagram is useful for determining the orientation and tilt of PV arrays and for modeling and designing automatic solar tracking systems that are used for both PV systems and solar concentrating systems.

The Earth revolves around the Sun along an elliptical orbit for a period of 365 days. From the Earth, it appears as if the Sun is moving from east to west from sunrise to sunset. But the diurnal motion is due to the daily rotation of the Earth about its own axis, for which day and night occur in all places of the planet. For an observer on the equatorial plane, the Sun will appear to go straight from east to west in a day. However, from the northern hemisphere, an observer shall see the Sun slightly on their southern sky (and their northern sky for the south hemisphere) due to the declination of the Earth with respect to the Sun. As such, the path of the Sun from the northern hemisphere is somewhat similar to Fig. 4.11.

There are two representations of the Sun path diagram—polar and Cartesian. Based on the scale of the altitude circles, the polar representation is again of three types—spherical, equidistant, and stereographic.

The Cartesian Sun path diagram is simply a two-dimensional graph, with the azimuth angle in the x-axis and the elevation angle in the y-axis. Therefore, the range of the x-axis is from 0° to 360°, while that of the y-axis is 0–90°. It is noteworthy

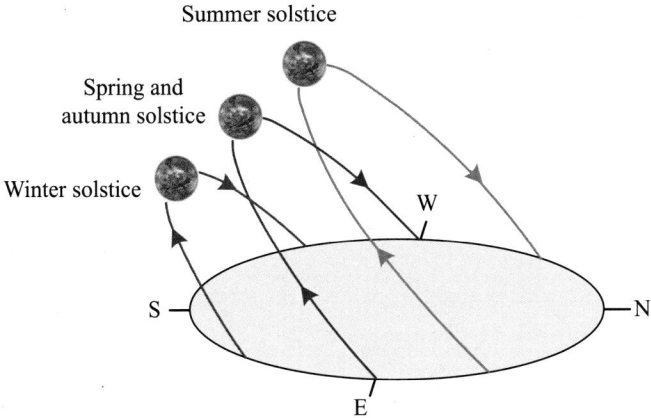

Fig. 4.11 The path of the Sun observed from the northern hemisphere. For the view from the southern hemisphere, simply switch the North with the South and the East with the West

that the time can also be read from the Sun path diagram from the angles in the x-axis, since each 15° corresponds to 1 h. Hence, each 90° advancement along the x-axis also implies the passage of 6 h.

For a specific location, the path of the Sun is fixed, although it varies with the time of the year. As Fig. 4.11 shows, the Sun is higher up in the sky during summer and lower during winter. Clearly, the elevation angle is higher in summer than in winter. Again, on a certain day, the elevation of the Sun from the horizon is the same at two instants—once while going up toward the solar noon and once while going down from the solar noon.

If the azimuth angle and the elevation angle of the Sun on each day of each year are plotted in the said graph, a Sun path chart will be obtained. By averaging the data points of the days in a month, twelve curves will be plotted on the graph. Again, since the path of the Sun mirrors itself half of the year, therefore, only six curves would suffice to completely describe the path of the Sun throughout a year. However, since the position of the Sun in the months of June and December is not exactly the opposite, therefore, one additional month is required to fully illustrate the paths of the Sun throughout the year. More accurately, the Sun path diagram reveals the solar window of a certain place, which provides a fair idea about the amount of insolation that the place receives on a yearly basis.

Figure 4.12 illustrates the Sun path diagram of Oregon, USA. The graph is drawn in the Cartesian system and portrays seven curves—each for the months December through June.

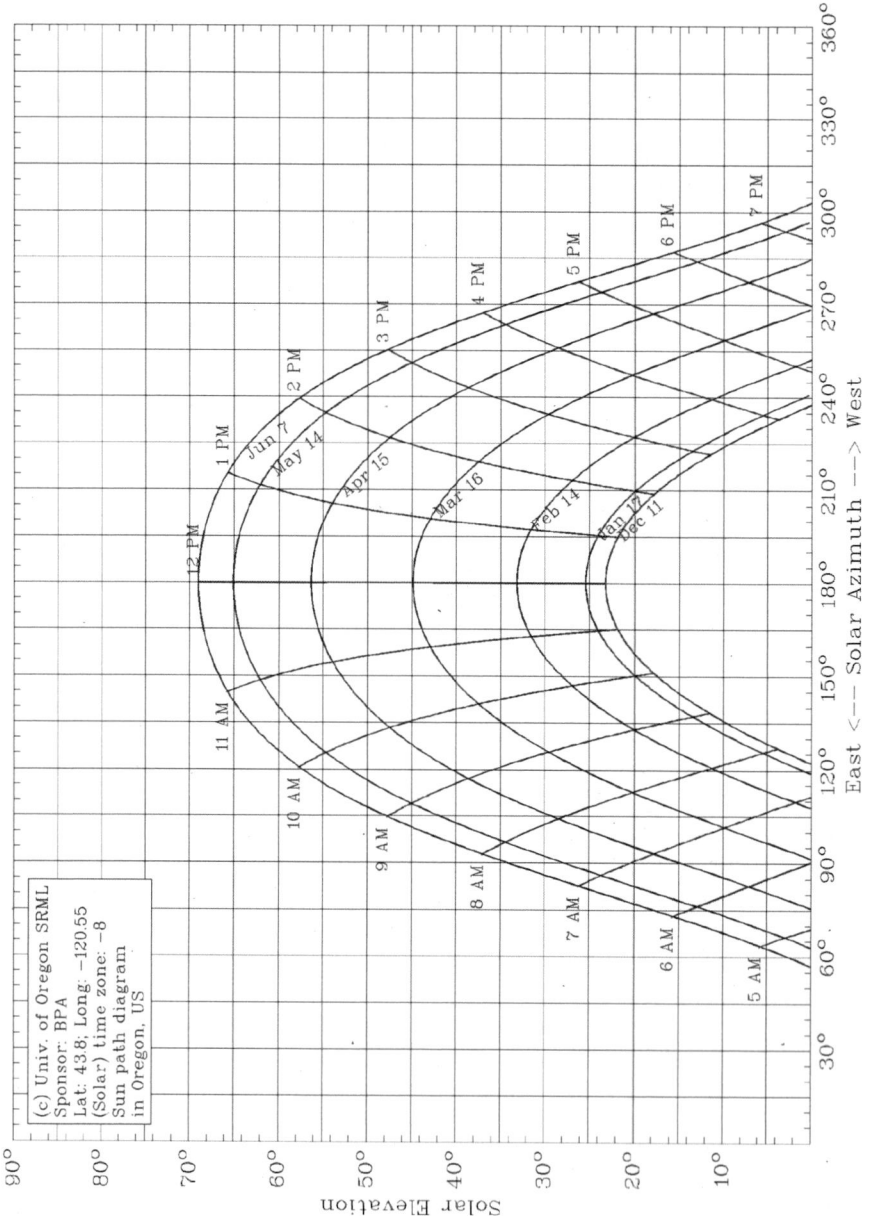

Fig. 4.12 The Sun path diagram in Oregon, USA. Generated by the free online Sun path chart program by the Solar Resource Monitoring Laboratory, University of Oregon, USA. The free tool is available at http://solardat.uoregon.edu/SunChartProgram.html

4.14   PV Module Orientation

The amount of sunlight befalling the Earth is the greatest around the equator and directly falls on the surface. As the latitudes increase on either side of the equator, the direct radiation on the surface diminishes. The best orientation for solar modules is to face the geographic or true south in the north hemisphere and the geographic or true north in the south hemisphere. For optimal production of solar energy, the solar module should be slightly tilted at an angle equal to the geographical latitude of the place. For instance, a solar module in Bangladesh should be tilted at an angle of 23.685° from the horizon toward the true south direction. So, the closer a place is to the equator, the less will be the angle of tilt; the closer it is to the poles, the more will be the angle of tilt facing toward the equator to attract maximum solar energy. As shown in Fig. 4.13, the tilt angle must be greater in winter than in summer due to the lower altitude of the Sun in winter than in summer (this can be comprehended from Fig. 4.11).

The tilt is an important consideration for setting up fixed solar modules. However, the tilt angle may be varied using an appropriate tracking mechanism that helps to change the tilt angle according to the movement of the Sun in the sky. Solar tracking systems are discussed in Sect. 5.6.8. The tilt of the solar array also depends on the base on which the module is to be installed. If the system is mounted on the ground, then the module tilt will be calculated based on the latitude of that location. But if the module is installed along the slope of an inclined rooftop, then the inclination angle of the rooftop must also be considered, and the total module tilt should be adjusted accordingly. In this case, the tilt angle of the module has to follow the existing angle of the inclined roof.

4.15   Effect of Tilt Angle and Orientation

The performance of a PV system is inextricably dependent on the tilt angle, δ, and orientation of the system. All non-concentrated solar applications must be tilted toward the radiation source (the Sun) to get a better output. Thus, it is crucial to

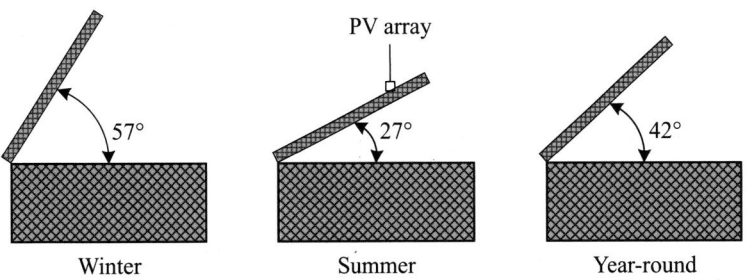

Fig. 4.13 Seasonal variation of tilt of solar PV modules. The tilt angle is higher in winter than in summer. The angles are based on using Method 1 of tilt angle calculation in Klamath Falls, Oregon, USA

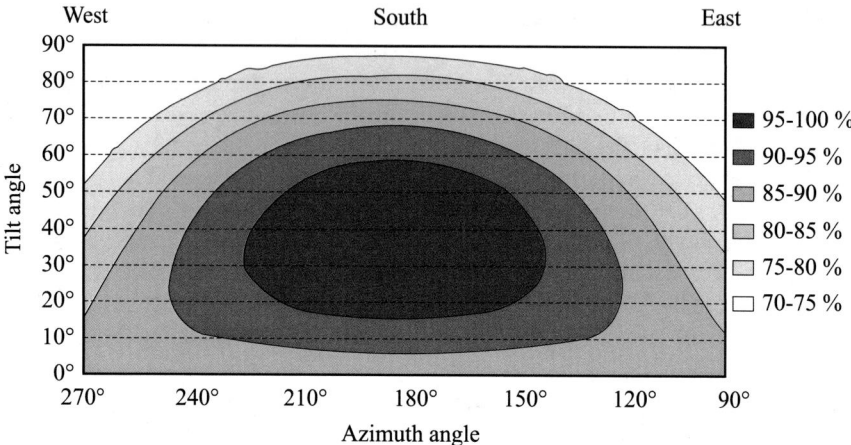

Fig. 4.14 The change in performance of a solar PV system in the northern hemisphere with the variation of orientation [5, 6]

find the best tilt angle for the solar modules used in the PV system. The tilt angle at which a solar PV system shows the best performance depends on the system location. It is seen that the solar modules are more tilted in winter than summer for obtaining maximum output. The tilt angle of the solar modules also depends on the solar azimuth angle, as shown in Fig. 4.14.

The tilt angle is ideally equal to the latitude. However, it changes according to the time of the day and also the seasons of the year. Therefore, the optimum tilt angle should be slightly adjusted accordingly. One thing to note here is that although the latitude is 0° at the equator, solar PV modules are not placed parallel to the ground in equatorial regions. To prevent the accumulation of dust on the modules and the resulting decrease in the power output, modules at the equator are also slightly tilted to maximize the energy harvest from the Sun.

Three methods can be described for calculating the tilt angle. The latitude of the city of Klamath Falls in Oregon is 42.2249° ≈ 42° North. Each method follows an example of the calculation of the tilt angle for Klamath Falls, Oregon.

Method 1: Seasonal Adjustment [7, 8]
During summer, the tilt angle is found by the following Eq. 4.10:

$$tilt_{summer} = Latitude - 15°. \tag{4.10}$$

During winter, the tilt angle is found by the following Eq. 4.11:

$$tilt_{winter} = Latitude + 15°. \tag{4.11}$$

And for the rest of the seasons (autumn and spring), the tilt angle is usually kept equal to the latitude [5]. According to this method, the tilt angle for Klamath Falls, Oregon, is 27° in summer, 57° in winter, and 42° for the rest of the seasons. These values have been depicted in the tilt angles in Fig. 4.13.

Method 2: More Precise Seasonal Adjustment [7]
This method is an advanced version of the previous one. The tilt angle, in this case, can be calculated using Eqs. 4.12 and 4.13.

$$tilt_{summer} = Latitude \times 0.9 - 23.5° \qquad (4.12)$$

$$tilt_{winter} = Latitude \times 0.9 + 29°. \qquad (4.13)$$

According to this method, the tilt angle for Klamath Falls, Oregon, is 14.3° in summer and 66.8° in winter. Similar to method 1, the tilt angle is kept the same as the latitude for the other seasons, i.e., 42°.

Method 3: Latitudinal Adjustment [9]
Unlike the other two methods, this method does not depend on the season. The tilt angle varies according to the value of the latitude of the place. The change in tilt angle is shown in Eqs. 4.14, 4.15, and 4.16:

- If $Latitude < 25°$,

$$tilt = Latitude \times 0.87. \qquad (4.14)$$

- If $25° < Latitude < 50°$,

$$tilt = Latitude \times 0.87 + 3.1°. \qquad (4.15)$$

- If $Latitude \geq 50°$,

$$tilt = 45°. \qquad (4.16)$$

According to this method, the tilt angle for Klamath Falls, Oregon, is $42 \times 0.87 + 3.1 = 39.64 \approx 40°$.

So, which method should we use and why? There is no definite answer to this, and the method for the calculation of the tilt angle is more of a personal choice to a designer. However, the most widely accepted method is the first method, mostly due to its simplicity and universality.

4.16 MATLAB Code to Determine GTI

The GTI values can be determined for each day of the year based on the GHI values at any given place with the help of a simple MATLAB code. MATLAB is a numeric computing platform that is used for performing calculations using programming, creating engineering models, and conducting simulated case studies. Although this book does not cover MATLAB programming, the readers may obtain a basic overview of its syntax with examples from Ref. [10].

The MATLAB code is provided in Listing 4.1. The code takes in several inputs and presents as the output the graphical illustration of the GTI over the year. The code provided in Listing 4.1 has been specified for Oregon, USA, whose output is shown in Fig. 4.15. The graph shows how the GTI values per day vary over the course of 1 year in Oregon, USA.

Listing 4.1 Determination of GTI

```
s=1.37; % solar constant; fixed value
phi=43.8 ; % insert location latitude
beta=44; % insert the tilt angle at which the GTI is calculated
rho=0.2; % surface albedo; depends on the surface; ranges from 0
     to 1
i=1;
for n=17:30:365
N(i)=n;
if(i==1)        G=1.68; % insert GHI in January
elseif(i==2)    G=2.61; % insert GHI in February
elseif(i==3)    G=3.61; % insert GHI in March
elseif(i==4)    G=5.32; % insert GHI in April
elseif(i==5)    G=6.67; % insert GHI in May
elseif(i==6)    G=7.87; % insert GHI in June
elseif(i==7)    G=7.39; % insert GHI in July
elseif(i==8)    G=6.73; % insert GHI in August
elseif(i==9)    G=5.14; % insert GHI in September
elseif(i==10)   G=3.59; % insert GHI in October
elseif(i==11)   G=1.89; % insert GHI in November
elseif(i==12)   G=1.36; % insert GHI in December
end
delta(i)=23.45*sin(3.14/180*(360/365*(N(i)-81)));
ws(i)=acos(-tan(3.14/180*phi)*tan(3.14/180*delta(i)));
Bo(i)=24/pi*s*(1+0.033*cos(3.14/180*(360*N(i))/365))*(cos(3.14
     /180*phi)*cos(3.14/180*delta(i))*sin(ws(i))+ws(i)*sin(3.14
     /180*phi)*sin(3.14/180*delta(i)));
kt(i)=G/Bo(i);
D(i)=G*(1.39-4.027*kt(i)+5.531*kt(i)^2-3.108*kt(i)^3);
Dbeta(i)=0.5*(1+cos(3.14/180*beta))*D(i);
Rbeta=0.5*(1-cos(3.14/180*beta))*rho*G;
B(i)=G-D(i);
wss(i)=acos(-tan((3.14/180)*(phi-beta))*tan(3.14/180*delta(i)));
wo(i)=min(ws(i),wss(i));
Bbeta(i)=B(i)*(cos((3.13/180)*(phi-beta))*cos(3.14/180*delta(i))*
     sin(wo)+ wo*sin((3.14/180)*(phi-beta))*sin(3/14/180*delta(i))
     )/(cos(3.14/180*phi)*cos(3.14/180*delta(i))*sin(ws)+ws*sin(3
     .14/180*phi)*sin(3.14/180*delta(i)));
Gbeta(i)=Dbeta(i)+Rbeta+Bbeta(i);
    i=i+1;
   end;
plot(N,Gbeta,'LineWidth',2.2)
grid on
xlabel('Number of days');
ylabel('GTI, kWh/m^2 per day');
title('GTI for Oregon, USA');
```

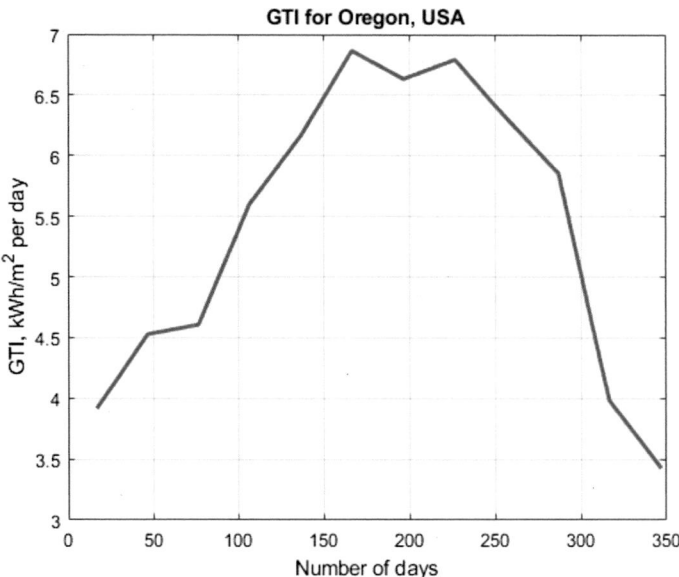

Fig. 4.15 The GTI for Oregon, USA, as found from the MATLAB code

To understand the code better, you may have a look at Ref. [1]. The line by line explanation of the code is provided below:

- For each location, `phi`, `beta`, `rho`, and the monthly GHI values need to be changed in the code. `S` is the solar constant, and it is always 1.37.
- The variable `i` indicates the number of months, and the variable `n` indicates the number of the days of the year, where `n=1` corresponds to 17th January.
- The GHI values used in the code from lines 8–19 can be obtained by any online free or paid resource. The NREL data, Meteonorm, etc., can be used to glean the GHI values. The values used in this code have been taken from Meteonorm.
- Equation 4.6 has been used in Line 21 in the code to determine the declination angle. The angle has been converted into radians by multiplying by $\frac{\pi}{180}$ for ease of calculation.
- Line 22 determines the sunrise hour angle `ws`.
- Line 23 determines the terrestrial radiation AM1.5 represented by `Bo`.
- Line 24 helps to calculate the clearness index `kt(i)`. The clearness index measures how clear the atmosphere is on a scale from 0 to 1 as the ratio of the surface radiation to the extraterrestrial radiation.
- Line 25 determines the diffused tilted irradiance `D(i)`.

- Dbeta in line 26 refers to the daily diffused tilted irradiance I_{dfi}, which can be expressed as shown in Eq. 4.17:

$$I_{dfi} = \left(\frac{1 + \cos \beta}{2} \right) \times I_d, \tag{4.17}$$

where I_d is the hourly diffused radiation on a horizontal surface, and β is the tilt angle measured from the horizon.
- Rbeta in line 27 refers to the daily reflective tilt irradiance I_r, which is expressed in terms of the horizontal irradiance I_H and surface albedo ρ as shown in Eq. 4.18:

$$I_r = I_H \times \rho \times \left(\frac{1 - \cos \beta}{2} \right). \tag{4.18}$$

- Line 28 equates the beam radiation on a horizontal surface as the global radiation on a horizontal surface minus the diffuse radiation on a horizontal surface.
- Line 29 determines the sunset hour angle wss.
- Line 30 takes the min of sunset and sunrise hour angle value.
- Bbeta in line 31 refers to the daily beam tilted irradiance.
- Line 32 translates to: Daily global tilted irradiance (GTI) = Daily diffused tilted irradiance + Daily reflective tilt irradiance + Daily beam tilted irradiance.

4.17   Site Assessment

Site assessment is an important consideration before the practical installation of a solar PV system. As a general rule of thumb, the selected place should have adequate exposure to an optimum amount of sunlight, be free from shading, have lesser sources of pollution, and have a suitable soil type. In practice, weather and shading analysis is necessary to bring the shades from surrounding constructions under consideration. Many engineering and designing tools are available to perform weather and shading analysis for assessing the site before installation. The site assessment is different based on the size and purpose of the solar system. For instance, a residential solar system has lesser considerations and options for the assessment compared to a grid-scale solar farm. For a solar home system, the important considerations are the orientation, tilt, and size of the solar array, not to mention the budget for the system. On the other hand, a large-scale solar farm must conduct a thorough site survey and consider the solar resource availability, feasibility analysis, the cost–benefit trade-off, the most efficient solar technology, system sizing, tracking mechanism, maintenance facilities, grid integration process, and synergic components, load analysis, energy efficiency, shade analysis, and so on. To summarize, a residential solar system only considers fulfilling the purpose of the owner in terms of harvesting solar energy and cutting down on bills, whereas a commercial project must also take into account the economic and technical feasibility of the project.

4.18 Conclusion

This chapter discussed the preliminary concepts of solar resources, which are important for the proper utilization of solar resources using PV technology. Having the ideas about the solar spectrum, the Earth's radiation equilibrium, and the global insolation helps to comprehend the nature of the energy gifted to us by the Sun on a daily basis. This chapter also explains the solar geometry, i.e., the solar angles, with illustrations. Understanding the path of the Sun helps us to read a Sun path diagram and extract valuable information about the position of the Sun at any time of the year. This chapter also discussed the tilt and orientation of PV modules and described the different types of tracking mechanisms. Lastly, the chapter briefs the reader about the site assessment for a residential and utility-scale PV system.

4.19 Exercise

1. Mention the ways in which incoming solar energy is dissipated while reaching the Earth.
2. What is a Sun path diagram? Explain its significance.
3. Define peak Sun hour and mention its significance.
4. What are the effects of tilt angle and orientation in the performance of a PC system?
5. Name the Sun angles that are measured for installing PV modules. Illustrate those angles with an appropriate diagram.
6. What is the declination angle of the Sun on 2nd May in a leap year? [Answer: 15.52°]
7. What are the key considerations in the site assessment of a solar PV system?

References

1. S. Abbas, M. Maleki, H. Hizam, C. Gomes, Estimation of hourly, daily and monthly global solar radiation on inclined surfaces: models re-visited. Energies **10**(1), 134 (2017)
2. Global Energy Budget. https://gpm.nasa.gov/education/lesson-plans/global-energy-budget
3. Air mass (2021). https://www.pveducation.org/pvcdrom/properties-of-sunlight/air-mass
4. F. Kasten, A.T. Young, Revised optical air mass tables and approximation formula. Appl. Opt. **28**(22), 4735–4738 (1989)
5. L. Rodríguez, Solar panel orientation: how using East–West structures improves the performance of your project (2021). https://ratedpower.com/blog/solar-panel-orientation/
6. Solarchitecture, Which is the optimal surface for solar installations? (2021). https://solarchitecture.ch/orientation-and-tilt/
7. D. De Rooij, Solar panel angle: how to calculate solar panel tilt angle? (2021). https://sinovoltaics.com/learning-center/system-design/solar-panel-angle-tilt-calculation/

8. Solar Reviews, Best solar panel angle: How do you find it—and does it matter? (2021). https://
www.solarreviews.com/blog/best-solar-panel-angle
9. Lighting Equipment Sales, How to calculate solar panel tilt angle (2021). http://
lightingequipmentsales.com/how-to-calculate-solar-panel-tilt-angle.html
10. E. Hossain, A Crash Course on MATLAB & Simulink for Electrical Engineers. (2021)

Solar System Components

5

5.1 Introduction

A solar photovoltaic (PV) system is much more than an array of navy blue or black modules. Despite being the most visible and the main part of the total system, the visible, navy blue or black, rectangular slabs only convert the light energy into electric energy. The rest of the components of the system perform the critical task of making the power usable for our appliances and transporting the power into our places of consumption. The components used in a solar PV system can be broadly described into 5 categories, such as:

1. Solar PV module
2. Battery
3. Charge controller
4. Inverter
5. Balance of system components, which includes the wiring, mounting, tracking, cooling, protection, and grounding

This chapter introduces a reader to all these components and describes how to choose the right size or type of these components for designing any solar PV system.

5.2 Solar Modules

PV modules are the essential parts of any PV system due to their vital role in producing energy to the load. All PV modules consist of a fundamental element, called a solar cell, responsible for converting solar irradiance into DC energy through a physical phenomenon called the photovoltaic effect. Once the solar cell is imposed to any solar irradiance, 0.5–0.68 V is produced between the two poles of the Silicon solar cell with an irradiance-proportioned current when the cell is

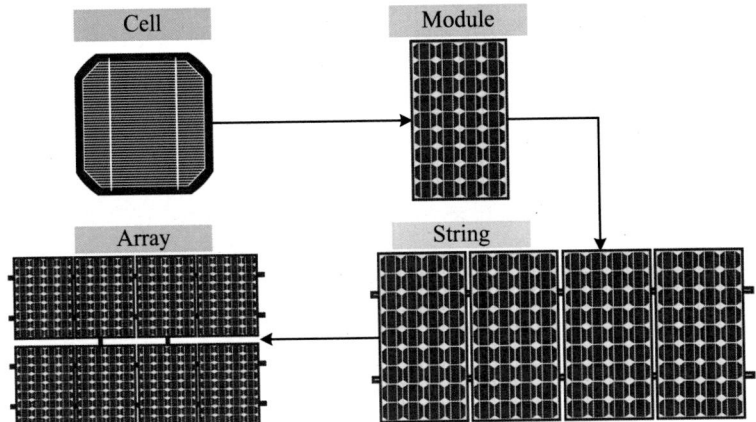

Fig. 5.1 Schematic diagram of solar array, string, module, and cell

put in a closed circuit. Figure 5.1 illustrates how several solar cells gather in a series configuration to create a solar module. The modules are connected in series to form a string. Then the strings are connected in parallel to form an array. This arrangement is made to satisfy the higher demands, starting from the scale of household appliances (several kWs) up to utility-scale projects measured in MW and even GW range.

5.2.1   Junction Box and Bypass Diodes

A junction box is a black plastic box attached to the upper edge of the backside of the PV modules. The metallic ends of the cell's busbars are connected to several diodes, called bypass diodes, in this box. During PV module operating phase, soft or hard shading occurs due to several causes, such as bird feces, trees, power or lighting poles, or even neighboring PV array with a short separation distance. The power productivity is significantly reduced once a partial shading is applied on one or multiple solar cells. In addition, power starts to dissipate as heat energy, leading to electrical and mechanical issues, as discussed earlier. A simulation process is conducted on a 72-cell, 310 W solar module from a well-known manufacturer to better understand the shading effect on the solar module. A continuous hard shading is applied on 1, 2, and 3 cells without bypass diodes, where the created shade covers 1.39%, 2.78%, and 4.17% of the module's total surface area.

The result shows that shading one of the 72 series-connected cells drops the output power more than 50%, while the energy yield does not exceed 50 W when three cells are shaded. Figures 5.2 and 5.3 show the change of I-V and P-V curves, respectively, based on the availability of shades.

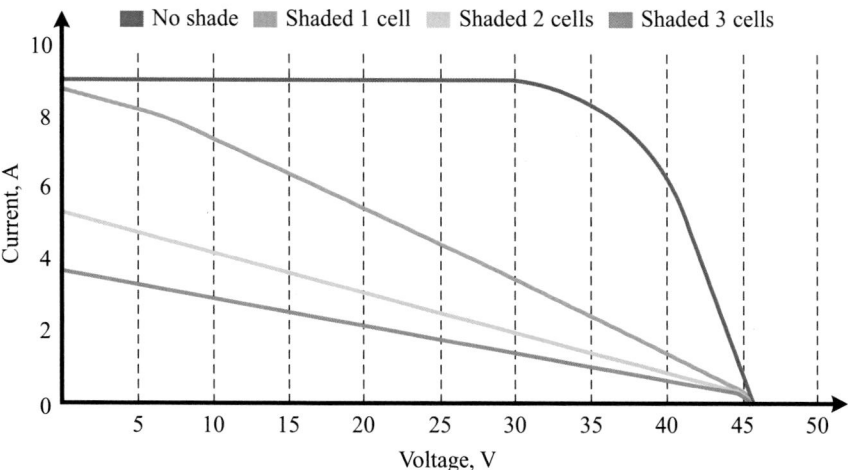

Fig. 5.2 Shading effect on I-V curve

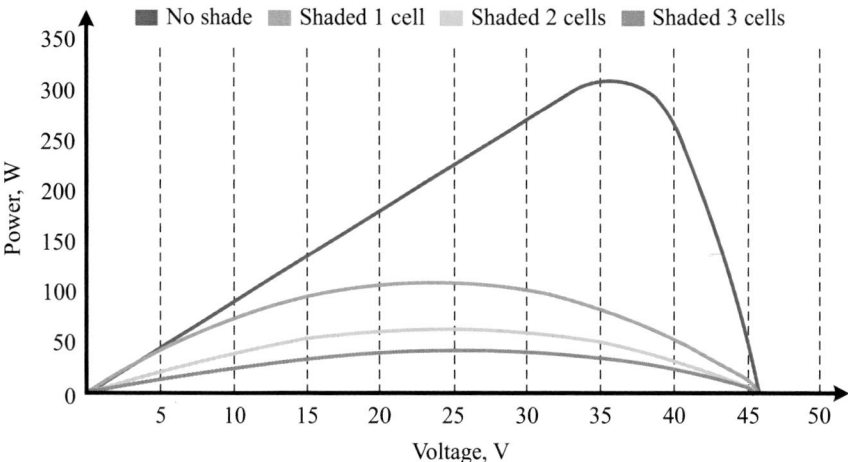

Fig. 5.3 Shading effect on P-V curve

Bypass diodes are connected with the series cells in several ways, for example: by the addition of overlapped bypass diodes and without overlapped bypass diodes as discussed below.

5.2.1.1 Addition of Overlapped Bypass Diodes
In order to add overlapped bypass diodes, 12 bypass diodes are added in parallel with the 72 cells at first, where a single cell can have more than one bypass diode. Then, they are connected in such a way that one bypass diode is parallel to each 7 series cell. The connection for overlapping bypass diodes is illustrated in Fig. 5.4.

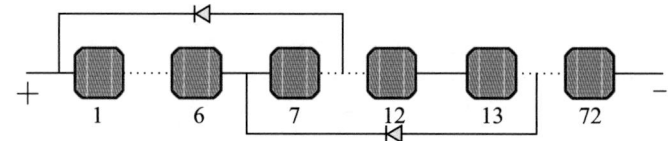

Fig. 5.4 Overlapped bypass diodes in solar cells

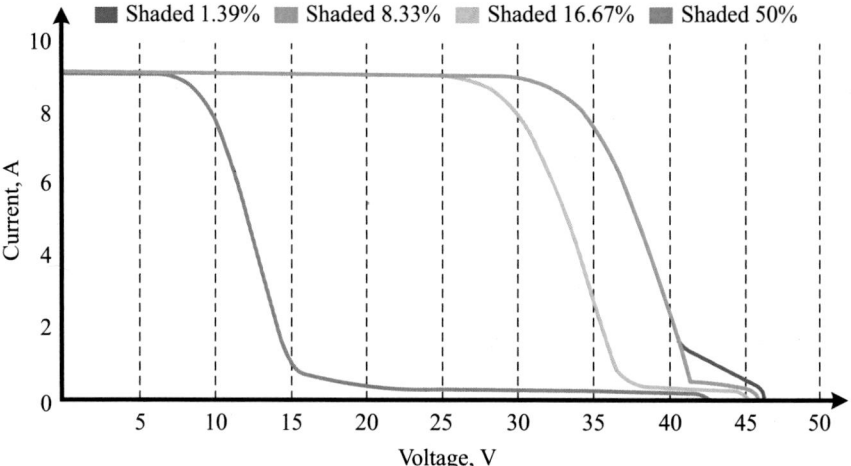

Fig. 5.5 Change in I-V curve for the presence of overlapped bypass diodes with shading effect

The simulation result of this connection of overlapped bypass diode shows a significant improvement in the energy yield, while shading effect occurs compared to no bypass diodes case. For instance, the maximum power point increased from 120 to 285 W with the overlapping bypass diode configuration. Figures 5.5 and 5.6 show the change of I-V and P-V curves due to the presence of overlapped bypass diodes, respectively. The figures describe a few shading cases and the essential performance of the bypass diodes. It is observable that the voltage drop increases and output power decreases with the rise of the shading effect.

5.2.1.2 Addition of Non-overlapped Bypass Diodes

In this configuration, the solar module is divided into 6 groups. Each group is formed by connecting 12 series cells (found by dividing 72 solar cells by the number of groups formed, i.e., 6) with a bypass diode in parallel. This type of connection does not include any overlapping of diodes 6, as shown in Fig. 5.7. The results showed a better performance than the two previously mentioned cases where shading 9 cells with a non-overlapped diode has a higher maximum power point than 6 shaded cells with overlapping diodes.

Figures 5.8 and 5.9 show the effect of shading on I-V and P-V curves without overlapped bypass diodes.

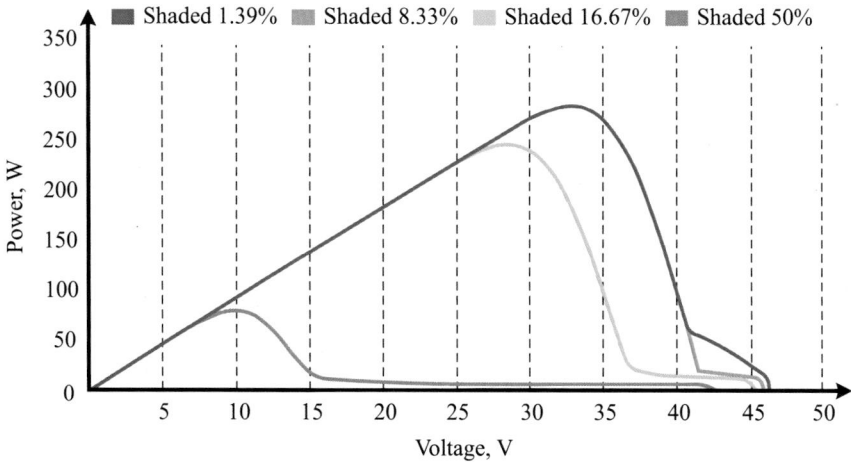

Fig. 5.6 Change in P-V curve for the presence of overlapped bypass diodes with shading effect

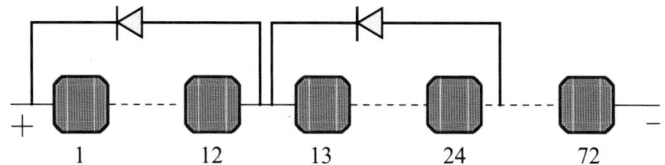

Fig. 5.7 Non-overlapped bypass diodes in solar cells

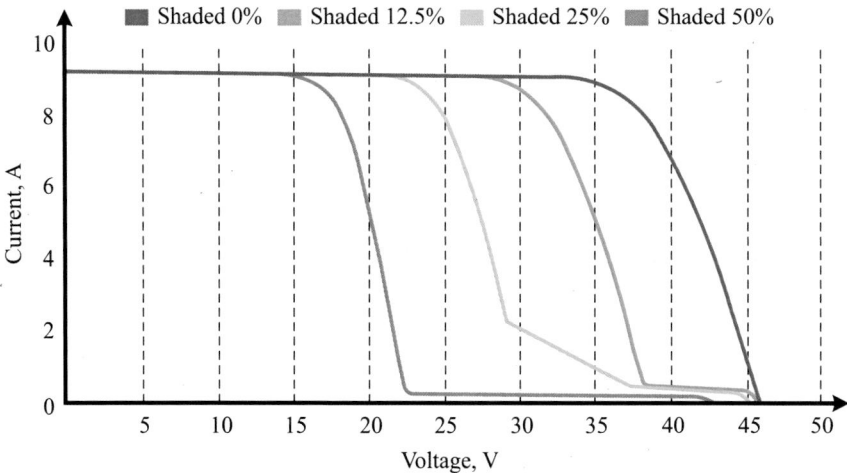

Fig. 5.8 Change in I-V curve in the absence of overlapped bypass diodes with shading effect

Usually, solar modules come with 3–6 bypass diodes. The electrical and thermal losses are dependent on the number of diodes inversely. With a higher number of bypass diodes, electrical and thermal losses reduce due to soft or hard shading [1].

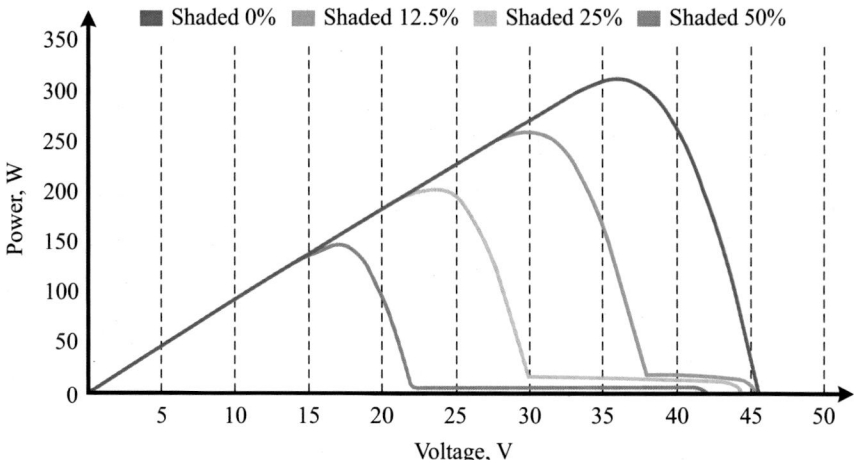

Fig. 5.9 Change in P-V curve in the absence of overlapped bypass diodes with shading effect

5.2.2   PV Module Terminals

Each module has two output terminals: positive and negative. These terminals connect the module with the load or with the neighboring modules to form a solar array. The terminal endings have male or female connectors to facilitate series connections. The MC4 Y connector is used for the parallel connection of the solar modules. The cables of solar modules are designed to handle the short-circuit current of the module. The wiring connection of series and parallel connection of the solar modules in terms of PV module terminals is described below.

5.2.2.1 Series Connection
Wiring PV modules in series increases the voltage. A series configuration of solar modules is facilitated by the use of male and female MC4 connectors, which makes the connection simpler. Each solar module contains a junction box at the backside, as discussed in Sect. 5.2.1. The junction box has two terminals, which are known as the PV module terminals. Wires are extended from both terminals of the junction box. To make a series connection, a module's positive terminal is connected to the negative terminal of another module, as shown in Fig. 5.10. The figure also shows the parallel connection of two modules using an MC4 Y connector. MC4 works as the medium of the connection, where the male and female connectors of MC4 and the terminals are connected based on the junction box labels. Several PV modules can be connected similarly.

5.2.2.2 Parallel Connection
Wiring PV modules in parallel will increase the amperage. The MC4 Y connectors are used for the parallel connection of PV modules. All the positive terminals are

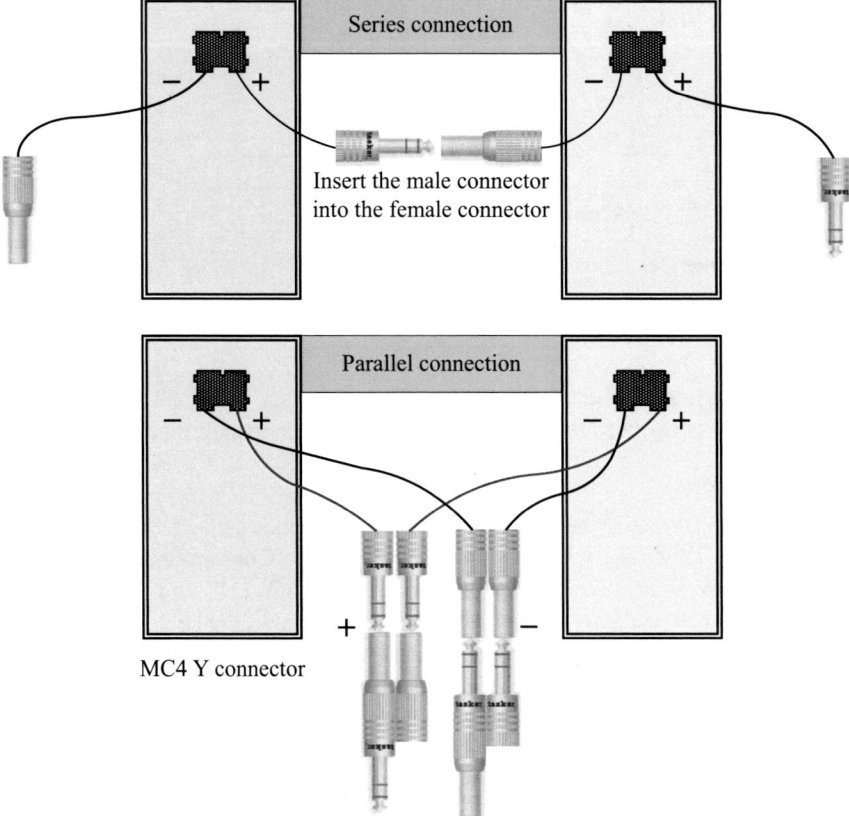

Fig. 5.10 Series and parallel connections of the MC4 for connecting PV modules

connected together in a parallel connection, whereas all the negative terminals are connected together separately. The positive terminals are joined together by one branch of MC4, and negative terminals by the other branch, as shown in Fig. 5.10.

5.2.3 Solar Module Types

Crystalline silicon PV modules can be distinguished into different categories based on several distinguishing factors, such as cell technology, cell size, backside availability, and the number of busbars.

5.2.3.1 Cell Technology

There are mainly two main types of C-Si, monocrystalline and polycrystalline. The monocrystalline manufacturing process is more energy-consuming than the polycrystalline due to the additional Czochralski process. This process is responsible for

Table 5.1 Comparison between monocrystalline and polycrystalline PV technologies

Factor	Monocrystalline	Polycrystalline
Color	Black	Dark blue
Si cell efficiency [2], %	26.7%	24.4%
Si module efficiency [2], %	24.4%	20.4%
Grains	None	Multiple
Annual PV production, GW_p, in 2020 [2]	120.6	23.3
Warranty	25–30 years	25–30 years
Average cell price [3]	0.143 $/Wp	0.114 $/Wp
Average module price [3]	0.234 $/Wp	0.213 $/Wp

decreasing the grains, which increases the efficiency of the cell. Table 5.1 illustrates the main differences between mono- and polycrystalline modules and cells.

5.2.3.2 Cell Size

Starting from 2014, solar manufacturers started to produce a new technology called half-cell solar modules, which is shown in Fig. 5.11. Conventional 60- and 72-solar cell-based solar modules were developed to be 120-cell- and 144-cell-based modules. The new technology is based on cutting the traditional cell into two equal parts without any mechanical crack or damage using state-of-the-art laser cutter technology. This process increases the mechanical stability of the cell since the area has been reduced to half, but thickness remains the same.

In the case of the 60-cell solar module, the 60 cells double to 120 cells and are distributed to the upper and lower sides of the solar modules. Each side has 60 cells connected in series. Both parts are parallelly connected to create the module's terminals where currents from both sides are combined, and the voltage remains the same. By applying this technology, the cell-generated current becomes 50% of the original current, while the voltage remains the same. The current is the same at the terminals of each cell, but it gets divided between the upper and lower parts. Therefore, the ohmic losses resulting from the current flows in the busbars resistance will be decreased, since $P = I^2R$. This leads to an increased overall efficiency due to minimizing the heat dissipation in the series resistor of the equivalent circuit.

The new arrangement will also maximize the benefits of the bypass diodes, which shifted from the upper side of the full-sized cell-based module to the center of the half-sized cell-based module. In the conventional module, the 60 cells are divided into three 20-cell groups. Each group is connected in parallel with a bypass diode that works and eliminates the group when one or multiple cells in the group get shaded. This means a 33% loss when one of the groups gets eliminated. In contrast, with the new configuration, the cells are divided into six 20-cell groups. Each one is connected to a bypass diode that eliminates 1/6th of the total capacity when one or multiple cells in a particular group get shaded. In other words, the new configurations give the module twice as resilience as what it has in the traditional case. That is why this technology is preferred in locations where designers are

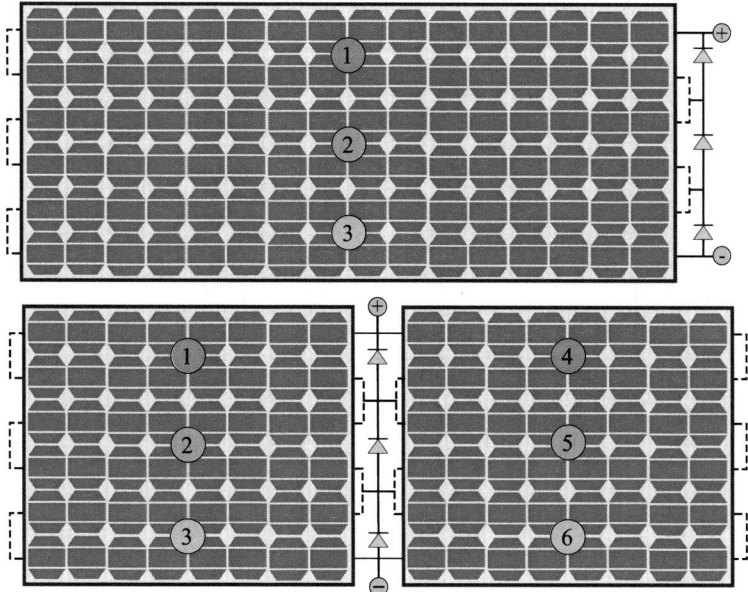

Fig. 5.11 Half-cell modules with bypass diodes [4]

expected some shades during the operating time. The market share of half-sized cell modules currently stands at 11%, which is prognosticated to reach 40% by the end of 2028.

5.2.3.3 Bifacial Modules

The surface of both the monofacial and bifacial modules is covered by an encapsulant. A layer of white backsheet is also present in monofacial modules, which blocks the solar cells from receiving any light reflected from the ground surface. In 2019, solar module manufacturers started to commercialize a pre-discovered technology in the laboratory called bifacial technology. Contrary to conventional monofacial technology-based solar modules, bifacial PV modules receive photons from both the rear and front sides. This increases the output values of current and power due to the utilization of the reflected and diffused light and the direct light beams. Thus, bifacial technology offers extra energy yield compared to the original energy yield of the monofacial technology. This relationship is expressed in Eq. 5.1.

$$Bifacial\ Gain = \frac{E_{bifacial} - E_{monofacial}}{E_{monofacial}}. \tag{5.1}$$

The equivalent circuit of the bifacial module is shown in Fig. 5.12, which looks like the monofacial equivalent circuit with an additional current source representing the rear side. This state-of-the-art structure is expected to give the bifacial modules

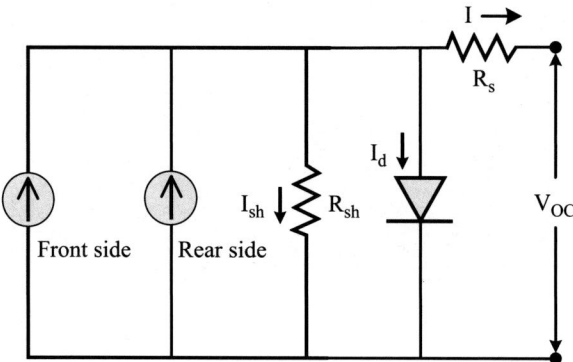

Fig. 5.12 Bifacial cell equivalent circuit

superiority over the monofacial in the next 6–7 years. Bifacial technology is being developed and tested in the labs. It has entered the commercialization phase with results creating promising expectations of exceeding the market share of the monofacial technology within 9 years. The bifacial gain depends on three factors mainly, which are as follows:

- **Mounting elevation:** The higher the module is installed, the more reflected irradiance reaches the rear side, and the lower is the self-shading effect.
- **Albedo:** A higher albedo translates to lesser irradiance absorbed by the ground and more reflected irradiance to be utilized on the rear side of the module.
- **Module orientation:** The orientation of the module depends on the tilt angle and latitude of a specific location. It also varies depending on the weather condition.

Maximizing Bifacial Modules Output

Bifacial modules have drawn the focus of many researchers in recent times. These modules are more effective than monofacial modules as they can generate power from both the front and rear sides. The front side of the module is in direct contact with solar irradiance. In contrast, the rear side generates energy by the reflected radiance.

A recent study [5] has focused on the optimization of the performance of bifacial modules based on the elevation, azimuth angle, optimum orientation, and tilt angle of the bifacial module. The global bifacial gain of ground mounted bifacial modules is less than 10% at a low albedo of 0.25. But, the gain can be increased to 30% at an albedo of 0.5 and raising the module 1 m above the ground. A higher installation height increases the chances of reflected irradiance to incident the rear side of the modules as well as reduces self-shading. The study also states that at an albedo of 0.5, ground mounted, vertical, east–west-facing, bifacial modules can outperform optimally tilted south–north-facing modules by up to 15% within the latitude of 30°.

5.2.3.4 Number of Busbars

Since the losses of solar modules are well-defined, decreasing the series resistor of the metallic contacts or busbars (BB) on the solar cell surface is always a possible approach toward increasing the efficiency. We know that a greater cross-section of conductors decreases its resistance. So, this reduction might be made by increasing the number of BBs (also called ribbons) or their cross-section (surface area). However, an increase in the surface area causes the active surface area of the PV cells to decrease, which falls under the ribbon's shades. Therefore, optimizing the number and surface area of the ribbons to achieve the best possible efficiency pushes to try a new concept called multi-busbars (MBB), where a cell is connected with the next one by 12 busbars instead of four or five.

A research was conducted using Si wafer along with the same parameters of the manufacturing processes, including texturing but with a different number of busbars. In the research, a group of cells was connected with 5 BBs and another one with 12 BBs. Researchers were able to reach 19.1–19.3% of the efficiencies with the 12 BB-cell-based modules, and the 5 BB-cell-based module efficiencies were limited to 18.6–18.7%, where the average powers were 278.07 W and 273.98 W, respectively. Adding 12 rounded ribbons instead of 5 flat ribbons leads to an increase in the efficiency and increase in the output power of the same number and surface area of cells [6].

5.2.4 Mismatching Problem

In many developing countries, solar PV projects are mainly initiated by non-governmental organizations and supported by donors who have extra PV modules from different projects. Many solar designers contact the local solar manufacturers to collect defective or class B solar modules with varieties of nominal power ratings to use them in humanitarian projects. In parallel, solar system designers might not obtain all required solar modules due to a shortage of supply or difficulties in transportation and shipping caused by natural disasters or military conflicts. All these causes lead to a problem called mismatching. It occurs when the designer has to build a solar system with solar modules that do not have the same electrical characteristics as the other system components. If a solar module has to be replaced during the project lifetime due to a failure, the same model may not be found again. The mismatch might also appear when a module has low performance due to soft or dark shading, which drives the whole system performance down. Therefore, some recommendations might reduce the damage of the mismatching problem, such as:

- **Replace the defective module with a higher-rated module without changing the product line:** Once the responsible project personnel is forced to replace a defective module(s), replacing it with a higher power-rated module is recommended because the original low-rated modules dominate the overall performance. In this case, the replaced higher-rated module will perform low to be synchronized with the other modules. In contrast, when replacing the defective module with a lower power-rated module, the original modules in the system will all be forced to lower their performance, leading to more losses. For example, if one module is defective in an array of 540 W rated modules, the defective module can be replaced with a 545 W rated solar module of the same product line.

- **Change the existing modules' arrangement:** Sometimes, it is a better option to change the system's configuration to mitigate the shading effect by a daily object in a particular period of the day. This problem might have a low impact when the shaded module is replaced with a higher power-rated module.
- **Current mismatch in a series configuration:** When solar modules with different MPP currents are connected in series, the overall voltage is found by adding the voltage of each module. The overall current is provided by the poor-performing module, which gives the lowest current of all. On the other hand, the extra current produced by the high-performing modules is dissipated as heat energy. Therefore, it is highly recommended to keep the deviation of the maximum power point not greater than 10% to avoid any damage done by the heat dissipation.
- **Voltage mismatch in a parallel configuration:** It occurs when two or more parallel modules have different maximum power point voltages. This problem leads the system to operate far from the MPP, which means high power losses. Similar to the current mismatch problem, the deviation of the maximum power point voltage, in this case, should not exceed 10%.
- **Change the entire string when half of the modules need to be replaced:** It is recommended to change the complete string when 50% of the modules are needed to be changed [7]. Another method to change the system is to split the big system into subsystems by maintaining the maximum power point voltage and current within 10% deviation, as mentioned previously.

5.3 Batteries

In a solar PV system, power is generated as long as sunlight is available. For this reason, systems in which solar energy is the only power generation source require a way in which the energy can be stored to use later or sometimes simultaneously with the system itself according to the demand. Because solar energy is intermittent in nature and its generation pattern does not match the load pattern, it requires an energy storage system (ESS). "Storage" refers to systems that absorb electricity, store it as another type of energy (chemical, thermal, or mechanical), and then release it when needed. There are different types of ESS available such as flywheel and pumped hydro power. But among all, electrochemical storage is suitable for a solar PV system and thermal storage (fluids) with CSP plants.

Before diving deep into the ESSs, let us first discuss the importance of storing energy. Without storage, solar energy should either be consumed immediately or put into the public grid, or else the energy will be lost. After sunset, the solar PV system cannot support the increased demand for electricity due to the decreasing generation of energy in the absence of sunlight. Moreover, the sunlight can be intermittent during cloudy days, resulting in little solar energy production where the demand at the load side remains the same. An ESS can aid in such cases by providing backup power when solar energy is limited or absent. It can charge up when the difference between generation and load is positive and discharge when the difference is negative, providing continuous energy during solar energy intermittency. Such short-term storage can also benefit the system by maintaining a constant output of the solar power plant during rapid variations in generation. Storage can also provide resilience. It can maintain critical infrastructure running to guarantee that important services, such as communications, are available at all times by providing backup power during electrical disruption.

In the form of a battery, electrochemical storage is by far the most prevalent method of storage for a solar PV system due to its improved energy density, compact size, and accessibility. Batteries are a vital component of any solar PV system, with a considerable impact on the PV system's cost, reliability, maintenance needs, and design. They produce electrical energy from the stored chemical energy, and the electricity can be used for any application when required.

Battery energy storage systems (BESS) can be used in both grid-tied and standalone types of solar PV systems. A BESS uses a rechargeable battery that stores energy from solar arrays. When surplus energy is available, the BESS acts as a load on the PV systems and functions as a source when it has to meet the demand of extra power by the load. Batteries are of two main types—primary or non-rechargeable batteries and secondary or rechargeable batteries. In all PV systems, rechargeable batteries are used.

Batteries in Standalone PV Systems

Standalone or off-grid PV systems are those that are not linked to the grid. Such systems use batteries for storing energy. Figure 5.13 shows a typical arrangement of an off-grid solar PV system with BESS. The fluctuating nature of the power generated by the PV systems necessitates the usage of batteries. The PV system directly feeds the energy to the load during the hours of sunlight, with additional energy being stored in the batteries for future use. Therefore, BESS allows the connected loads to run when the solar modules do not provide adequate electricity. A charge controller or charge regulator between the solar PV modules and the batteries ensures that the solar array's maximum power is directed toward charging the batteries without causing any damage due to overcharging. It also prevents the battery from over-discharging. When the battery is low on charge, the controller automatically detaches the battery from its electrical loads. The figure also shows an inverter for connecting the charge controllers with the AC loads and for regulating the variable power output. Simple standalone DC systems do not require any inverter. However, all forms of AC systems use additional components such as inverters, fuses, and rated wires for connection.

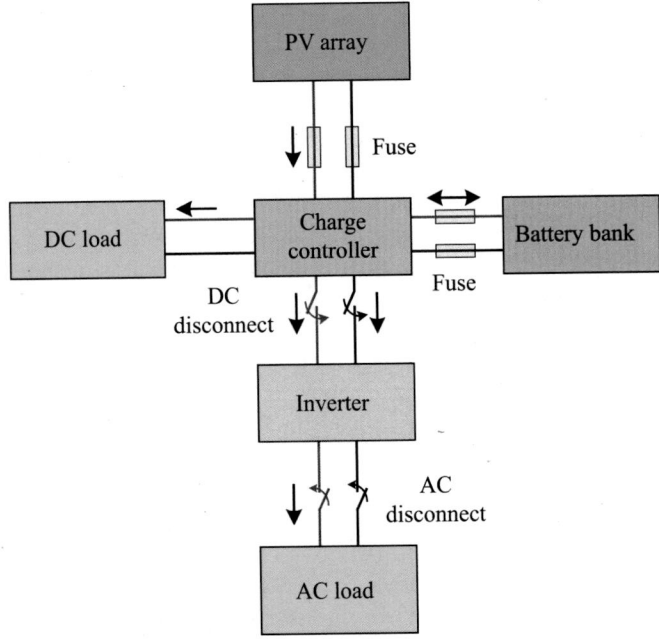

Fig. 5.13 Off-grid solar PV system with battery storage. For hybrid PV systems (grid-tied systems with battery), an additional path exists between the grid and the inverter. In the case of grid-tied systems (without battery), the inverter output is fed to the utility grid alongside the AC loads, but the battery bank is absent

Batteries in Hybrid PV Systems

A hybrid solar PV system is a grid-tied system with a BESS for storing backup power for an unexpected grid power outage. This system allows the battery to be charged by either grid power or solar power. The switching device connects the solar PV generation to the electricity grid. Charging the battery occurs when the solar PV system produces the most power, and discharging occurs when the solar PV system produces no or less power or when the load demand is high. If the demonstrated system in Fig. 5.13 is to be modified as a grid-tied system, the AC power output from the inverter is to be fed to the utility grid as well.

5.3.1   Common Rechargeable Batteries

Rechargeable batteries are those types of batteries that may be used as a load or a source as needed by interchanging the electrochemical reaction. The life cycle of such batteries varies based on their chemistry. Rechargeable batteries come in many different shapes and sizes according to their application. Some major types of rechargeable batteries are as follows:

- Lead-acid batteries
- Advanced Lead-acid batteries
- Alkaline (Nickel) batteries
- Lithium batteries

This section provides a basic overview of the four types of batteries mentioned above. To learn more about other battery types in detail, the readers may refer to Ref. [8–10].

5.3.1.1 Lead-Acid Batteries

The oldest rechargeable batteries are lead-acid (LA) batteries, with the widest variety of applications in both small- and large-scale power systems. They come with the advantages of low cost and negligible maintenance but have moderate efficiency and low energy density. These batteries consist of the anode made of lead dioxide and the cathode made of sponge lead, and the solution of sulfuric acid acts as the electrolyte. This classic design is also known as *flooded lead-acid* battery and requires high monitoring of the electrolytes and replacement of water periodically. Variations in the acid concentration within the battery can cause non-uniform utilization from the top to the bottom of the electrodes resulting in a reduced life cycle.

Another type of LA battery design is the *sealed design* that is also called *valve-regulated lead-acid (VRLA)* battery. It is a type of LA battery in which the electrolyte is formed into a gel. Even though this design has a number of advantages over the flooded LA design, it still requires additional charging requirements and meticulous thermal control [11]. Sealed LA batteries are of two types—gel type and AGM type.

An absorbent glass mat (AGM) is used between firmly packed flat plates in an AGM battery. AGM batteries were designed primarily for high-current (short discharge period) applications. They contain little acid, meaning they are extremely vulnerable to water loss, which often occurs at high temperatures. Because there is room for expansion within the AGM, they are not easily frozen solid.

On the other hand, sulfuric acid is combined with finely divided silica to make a thick paste or gel in LA gel batteries. The gel dries throughout the process of construction of the battery and has microscopic cracks that permit the cathode and anodes to exchange gas that is needed for the recombination process. In comparison to AGM batteries, the gel is better at conducting heat from the plates to the cell walls, allowing the overcharging heat to be dissipated more efficiently.

Components

In a LA battery construction, the most important components are the electrodes, container, grids, separator, electrolyte, and top lid. The container has to give protection to the electrolytes and electrodes and at the same time be resistant to sulfuric acid. The container is sealed tightly with a top lid and is typically made of lead-coated wood, glass, ebonite, hard rubber, ceramic materials, and forged plastic. The top lid has a hole in it for vent plugs and posts. The bottom of the container is ribbed to prevent a short circuit between the electrodes by holding them in place.

Two types of plates are used mainly in LA batteries—pasted flat plates and tubular plates. The former is the most common form of plates and is mostly used to construct the cathode, while the latter is used for anodes and has a longer life cycle. The separators are positioned between the plates to provide insulation and are made of coated leadwood, porous rubber, and glass fiber [8].

5.3.1.2 Advanced Lead-Acid Batteries

Although VRLA did not improve the overall performance of the LA cell, it did overcome the major problem of high maintenance. Numerous researches have been conducted to overcome the poor life cycle of the LA battery. Integrating carbon into the cathode of VRLA opened a new path for these batteries. It dramatically improved the life cycle and enabled VRLA batteries to enter new application areas. Operating at a partial state of charge (pSoC) can quickly reduce the life of a battery. The addition of carbon allowed VRLA to operate in pSoC for extended periods. The Commonwealth Scientific and Industrial Research Organization (CSIRO) developed the "UltraBattery" technology, which notably increased high-rate pSoC operation and improved power handling, efficiency, and endurance [12].

Components

The anode of the UltraBattery is fabricated with PbO_2, while porous Pb materials are used to make the cathode. Carbon-based counterpart is used in parallel with the cathode forming an asymmetric supercapacitor (also known as ultracapacitor). This is what makes the UltraBattery "ultra"İ as it integrates a supercapacitor with the LA battery. The anode and the supercapacitor can be incorporated into one unit since they have a common composition. It allows the supercapacitor to be internally connected with the cathodes in parallel, and thus both the electrodes end up sharing the same anode.

Supercapacitor allows the UltraBattery to charge or discharge at a very high rate while buffering the current to protect the components by providing high power. It is also responsible for prolonging the service life of the battery [11].

5.3.1.3 Nickel-Based Batteries

Batteries that use nickel as a cathode are commonly known as nickel-based batteries. Nickel-Cadmium was the first commercially available nickel electrode-based battery where Cd is used as the anode. Ni-Cd battery has better energy and power density as well as a higher number of cycles compared to LA batteries. However, the energy density that Ni-Cd batteries produce is lower than that of Li-ion batteries. Some of the other disadvantages include complex charging, a high self-discharge rate, and poor performance at high temperatures. In addition, cadmium being a toxic substance, Ni-Cd batteries are considered harmful to the environment. So, when used in a PV system, the disposal at the end of its life has to be done carefully. Ni-Cd battery also suffers from the "memory effect," which refers to the phenomenon that causes the batteries to hold less charge. As a result, the batteries gradually lose their capacity due to repeated recharge after being only discharged partially.

There exists another type of nickel-based battery known as Nickel-Metal Hydride cell. In this cell, the anode is made of hydride material. Compared to Ni-Cd, it is less vulnerable to the "memory effect" but more sensitive to overcharge. In addition, because it generates more heat, a complex algorithm of charge is required for Ni-MH batteries [8].

Components

In a Ni-Cd battery, alkaline (KOH) is used as the electrode. The anode and cathode are made of Nickel oxy-hydroxide (NiOOH) and cadmium metal, respectively. A nylon divider separates anode and cathodes, blocking any kind of direct charge transfer.

There are two variations of Ni-Cd batteries based on construction—sealed Ni-Cd battery and vented Ni-Cd battery. A metal case seals structure in a sealed Ni-Cd battery, and therefore, no gas leakage is allowed unless a fault occurs. Vented Ni-Cd has a similar operating principle, but they permit gas leakage through a low pressured valve during overcharging or high-speed discharging. Ni-MH batteries are almost as same as Ni-Cd batteries, just that Hydrogen is used as cathode instead of Cadmium. Substituting Cd with Hydrogen gives it some additional benefits, such as making it eco-friendly, increased battery capacity, and higher energy density. However, during fast charging and discharging, they produce heat and are prone to self-discharge, a phenomenon that reduces the stored charge without any connection between the electrodes or any other external circuit.

5.3.1.4 Lithium Batteries

Lithium batteries are the most popular and advanced form of battery. For portable electronic devices, it is a familiar component for providing power. It has many advantages in its application to PV systems compared to LA batteries. For instance, they have a higher life cycle and energy density than LA batteries. Lithium batteries are also eco-friendly [13]. The four types of Li batteries are Li-ion, Li-Metal, Li-Polymer, and solid electrolyte Lithium batteries. Lithium anode is used for all types of Lithium batteries, but the cathode material differs according to the characteristics required for different applications.

Components—Lithium-Metal Batteries

Theoretically, the energy content of the Li-Metal battery is higher, but it has several safety issues. In the early stage, the cathode was TiS_2, but the design resulted up in an explosion. Scientists then tried to insert different materials within layers of lithium, aiming for a safe and efficient design. The reactivity was reduced by inserting graphite into lithium. These gradual developments in Li-metal batteries progressively gave birth to the invention of the Li-ion battery, which later won the Nobel Prize in 2019. Other than the explosion problem, lithium-metal batteries also suffer from dendrite formation. Dendrites are formed at the anode, and they resemble small spike-like projections that can create a short circuit in the battery by tearing the separators between anode and cathode. Researchers are working to mitigate these issues, and now, lithium-metal batteries are set for a comeback [9].

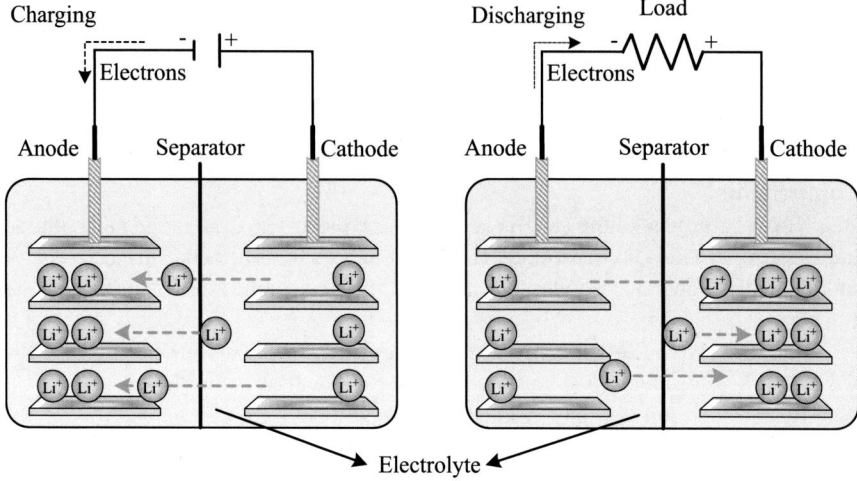

Fig. 5.14 The schematic of the charging and discharging processes in a Li-ion battery. The arrow represents the flow of electrons. By convention, the flow of current is in the opposite direction of that of the electrons

Components—Lithium-Ion Batteries

Li-ion batteries are the most popular batteries at present due to their superior energy-to-weight ratio, low self-discharge rate, high open-circuit voltage, no memory effect, and high energy density [14]. They mainly consist of an anode made of lithiated metal oxide (typically $LiCoO_2$), layered graphitic carbon as cathode, and a separator. Lithium salt ($LiPF_6$) is used as the electrolyte. The components of a Li-ion battery can be observed from Fig. 5.14. Depending on the cathode material, Li-ion batteries can be classified as lithium cobalt oxide (LCO), lithium manganese oxide (LMO), lithium iron phosphate (LFP), lithium nickel manganese cobalt oxide (NMC), lithium titanate (LTO), and lithium nickel cobalt aluminum oxide (NCA)[9].

Components—Lithium-Polymer Batteries

The construction and operation principle of Li-polymer batteries are similar to Li-ion batteries, except for the electrolyte material. In the early stage, a dry solid polymer was used as an electrolyte. But it had a high internal resistance, resulting in poor conductivity and making the battery unsuitable for high current applications. At present, gel electrolyte is mixed with the dry polymer in the latest version of Li-Po batteries. This resolves the poor conductivity problem and provides a high discharge rate to the battery, making it advantageous for high current applications.

A Li-Po battery consists of an anode and cathode, separator layer, and electrolyte. The anode consists of three parts: graphite or acetylene black serving as a conductive additive, transition metal oxides such as $LiMn_2O_4$ or $LiCoO_2$, and polyvinylidene fluoride $[-(C_2H_2F_2)_n-]$ functioning as polymer binder. The cathode has the same two parts other than the lithium-metal oxide, which is replaced by carbon. For electrolyte, conductive lithium salt such as $LiFP_6$ is used, and polypropylene $[(C_3H_6)_n]$ film performs the role of the separator layer[9].

Components—Solid Electrolyte Lithium Batteries

Although high voltage cathodes used in Li batteries increase the energy density, the practical use of it is impeded by innate problems of traditional batteries such as cathode producing dissoluble transition metals, liquid electrolyte producing narrow electrochemical window, and poor safety scheme of both Li and electrolyte. During high voltage applications, the carbonate-based electrolyte experiences continuous oxidative decomposition and makes solid non-passivating electrolyte interphases (SEI) films, which in turn reduces coulombic efficiency and life cycle. These SEI films also cause chemical etching and dissolution of metal ions that increase the thickness of SEI films and decrease the capacity available in cathodes. Furthermore, these electrolytes contain flammable ethers and esters that can cause safety issues while operating beyond their stable voltage limit. The use of Li-metal anodes is avoided due to Li dendrite growth, and commercially used graphite anodes have lowered operating voltage and capacity.

Solid electrolytes have a higher voltage range (above 5 V), and therefore, the cathodic current cannot decompose it. Moreover, the dissolution of transition metal into the electrolyte is much less of a concern with solid electrolytes. They are also less flammable comparing with carbonate electrolytes and last but not least, due to the mechanical robustness property, dendrites are less likely to form as Li metals are compatible with many solid electrolytes [15].

Based on the major categories mentioned above, LA, Ni-Cd, Ni-MH, and Li-ion batteries can be considered the most used. A comparative chart among these four batteries is presented in Table 5.2 to understand how batteries are distinguished based on their technical parameters.

5.3.2 Battery Parameters

There are several essential parameters in a battery datasheet that are important to understand their characteristics. The following subsections will aid the understanding of some of the important parameters.

5.3.2.1 Depth of Discharge

The depth of discharge (DoD) indicates the discharge amount in percentage corresponding to the battery's overall capacity. Typically, it can be measured by using Eq. 5.2 [20]. For example, if a battery holds 1.25 kWh of energy and 1.20 kWh has been discharged, the DoD would be approximately $(1.20 \div 1.25) \times 100\% = 96\%$. This can be explained better by the battery percentage on a phone or laptop. The battery percentage indicates how much charge is left in the device; the lower the percentage, the greater is the depth of its discharge, and the less charge the battery holds. For better understanding, see Fig. 5.16.

Table 5.2 Comparison of different types of battery cells used in solar PV system [16–18]

Parameter	LA	Ni-Cd	Ni-MH	Li-ion
Anode	Pb	Cd	$MH(LaNi_5H_6)$	LiC_6
Cathode	PbO_2	NiOOH	NiOOH	$LiCoO_2$
Electrolyte	Aqueous H_2SO_4	Aqueous KOH	Aqueous KOH	$LiPF_6$
Theoretical specific energy density, Wh/kg	170	220	220	410
Working temperature, °C	−20 to +50	−40 to +45	−20 to +45	−30 to +80
Nominal cell voltage [19] , V	2.0	1.2	1.2	3.7
Charge/discharge efficiency, %	70–92	70–90	66	99.9
Cycle number, 100% DOD	>200	>1000	>1000	<1000
Calendar life, years	2–8	3–10	2–5	2–3
Overcharge tolerance	High	Moderate	Low	Low
Safety requirement	Thermally stable	Thermally stable, fuse protection	Thermally stable, fuse protection	Protection circuit mandatory
Self-discharge per month (room temp)	5%	20%	30%	>5%; Protection circuit consumes 3%/month
Maintenance requirements	3–6 months	Full discharge every 90 days when in full use	Full discharge every 90 days when in full use	Maintenance free
Toxicity	Very high	Very high	Low	Low
Coulombic efficiency	~90%	~70% slow charge; ~90% fast charge	~70% slow charge, ~90% fast charge	99%
Cost	Low	Moderate	Moderate	High

$$DoD = \frac{Amount\ of\ charge\ removed\ from\ the\ battery,\ Q_d}{Maximum\ available\ amount\ of\ charge,\ C} \times 100\%. \quad (5.2)$$

In a solar PV system, the values of DoD are of two types: the average daily DoD and maximum allowable DoD. The average daily DoD illustrates the daily average discharge level compared with the full battery capacity. The average daily DoD and battery capacity are inversely related to each other. Systems designed for longer backup periods, in other words, for more capacity, will end up with a lower average daily DoD. In the case of seasonal loads such as air conditioners or water heaters, the

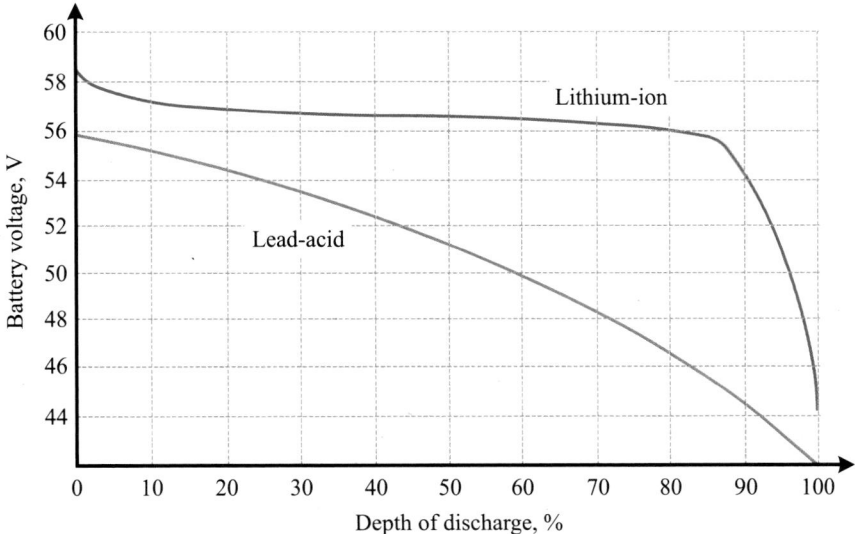

Fig. 5.15 Comparative discharge curve of LA and Li-ion battery. The battery voltage decreases with increased DOD

daily discharge level will also change. Ambient temperature also has effects over the average daily DoD. When the ambient temperature reduces, it reduces the battery capacity as well. In such occurrence, the daily DoD will increase. To prolong the battery life, the daily DoD should be kept low, i.e., the battery should not be drained too much.

The maximum allowable DoD can be denoted as the highest percentage of the rated capacity that can be extracted from a battery. Since the battery capacity is also dependent on maximum allowable DoD, this value must be fixed when designing the PV system. In standalone PV systems, the tolerable DoD limit at a definite discharge rate is dictated by the LVD (low voltage disconnect) set point of the charge controller. Excessive unexpected load, low temperatures, and seasonal shortage in the sunshine are some of the key factors that trigger this limit. Figure 5.15 shows the discharge curve of Lead-acid and Li-ion batteries.

5.3.2.2 State of Charge

The state of charge (SoC) implies the state of charge of a battery, i.e., it expresses the remaining charge of a battery, expressed in percentage. For example, 100% SoC implies a fully charged battery, and 50% implies a half-charged battery. So, the state of the battery at any point in its life span is displayed by the SoC of the battery. It helps to control the system accurately and thus increases system reliability. SoC is basically the opposite of DoD and is directly measured by the battery percentage of the phone or laptop. Figure 5.16 illustrates the relationship between DoD and SoC more clearly. SoC can be calculated from Eq. 5.3 [20]. However, to determine the

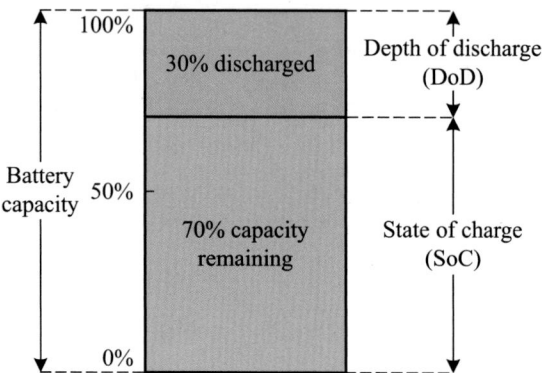

Fig. 5.16 The inverse relationship between the SoC and the DoD of a battery

actual available amount of charge, the discharge amount is required. In other words, the available amount of charge = initial amount of charge at 100% SoC—charge depleted during discharge.

$$SoC = \frac{Amount\ of\ charge\ available,\ Q}{Maximum\ available\ amount\ of\ charge,\ C} \times 100\%. \qquad (5.3)$$

As illustrated in Fig. 5.16, it can also be implied that

$$SoC + DoD = 100\%. \qquad (5.4)$$

5.3.2.3 Discharge Rate and Charge Rate

A discharge rate measures the rate at which a battery is discharged with respect to its maximum capacity. It is often expressed as C-rate. For example, for a battery of 1 Ah, the battery's discharge rate is 1 C, which means the battery would provide 1 A current for 1 h. For the same battery, a 0.2 C rating means that it would provide 0.2 A or 200 mA current for 5 h, and a 2 C rating indicates that it would provide 2 A for 0.5 h or 30 min. On the contrary, the charging rate can be expressed as the amount of charge added to the battery per unit of time. It is expressed as the number of hours as shown in Eq. 5.5:

$$Rate = \frac{Capacity,\ Ah}{Current,\ A}. \qquad (5.5)$$

There is an upper limit for the anode on the number of lithium ions they can transport per unit of time in a Lithium battery. This is because during the charging process, letting excessive current enter the battery forcefully causes "lithium plating," a phenomenon in which a lithium-metal layer is formed when

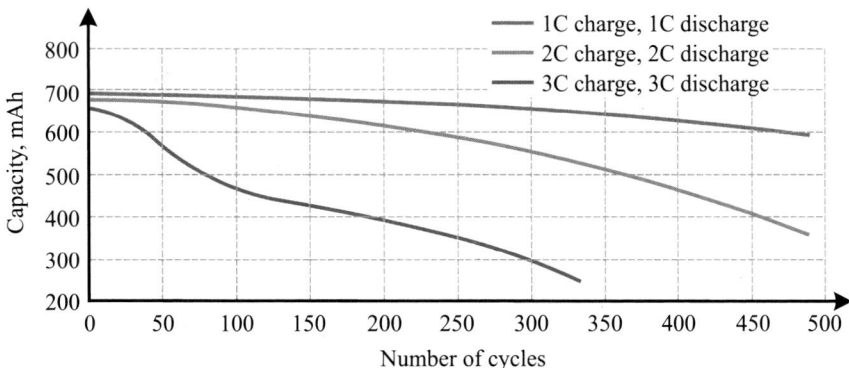

Fig. 5.17 Battery capacity versus the number of cycle curves under different charging and discharging rates for lithium-ion batteries [21]

excess lithium ions accumulate on the surface of the electrode. This causes internal impedance growth as well as capacity loss.

From Fig. 5.17, it can be seen that, at increasing C-rate, the capacity and life cycle of the battery reduce significantly, as the active chemical transformation cannot keep up with the current drawn.

5.3.2.4 Self-Discharge Rate

The self-discharge rate is the loss of electrical capacity with respect to time when the battery sits idly, owing to internal electrochemical processes. A battery's self-discharge rate is proportional to the rise of ambient temperature. It also depends on factors such as the battery type, charging current, and SoC. It is typically identified as a percentage of capacity loss each month from a completely charged battery. The duration for which a battery stores energy without losing its capacity is called its shelf time. Generally, it is preferred that the self-discharge rate remains low as it increases the shelf life of the battery (between 1 and 4% per month at 15–20 °C) [22].

5.3.2.5 Cycle Life

The cycle life refers to the number of full charge–discharge cycles that a battery can handle before its nominal capacity reduces to below 80% of its rated initial capacity. A battery can be used after its rated specified life cycle, but the capacity will be lower. The cycle life of batteries is affected by DoD, discharge rate, and temperature. Discharging the battery regularly at a lower DoD will leave it with more useful cycles than if it were to be frequently drained to its maximum DoD. Figure 5.18 demonstrates the relationship of the battery capacity with the number of cycles at different DoD values for an LA battery.

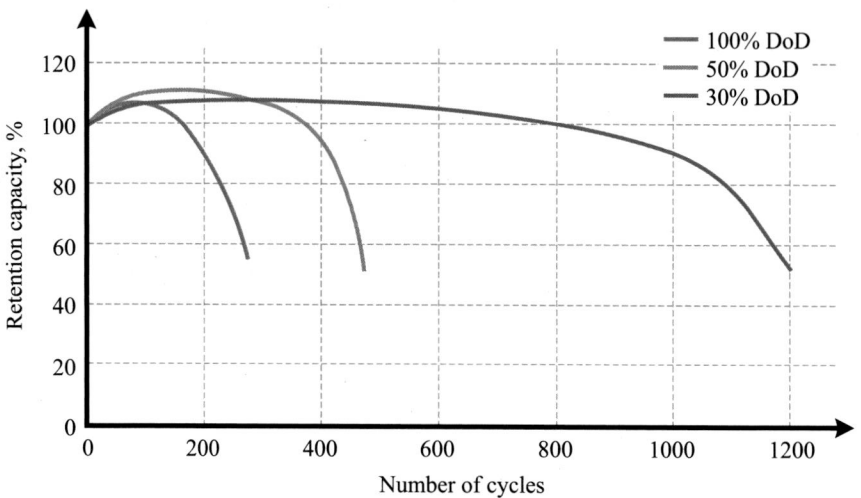

Fig. 5.18 Relationship between battery capacity, depth of discharge, and cycle life [23]

5.3.2.6 Capacity

In a standalone solar PV system, different types of batteries are used, and they come with different types of capacity. The battery capacity measures the charge contained by a battery in Ampere-hour (Ah). This rated capacity is represented as the maximum Ah deliverable by a fully charged battery under specific conditions, such as:

- The cutoff voltage or the voltage at which the battery is discharged
- The current at which the discharge occurs
- The battery temperature

The discharge rate has to be diligently noted along with any capacity. For example, an ideal battery rated at 100 Ah indicates that the battery will discharge within 10 h at a rate of 10 A, or within 1 h at a rate of 100 A, or within 100 h at a rate of 1 A.

For PV system design and battery sizing, capacity plays an important role in determining the days of autonomy. The days of autonomy for a battery system are the number of days the battery can serve the system without charge. For a large system, a larger capacity of the batteries is expected for more days of autonomy. Therefore, the nominal battery capacity should be chosen so that there are sufficient days of autonomy of the PV system. In addition, since battery capabilities are affected by cold temperature and by the value of the cutoff voltage, i.e., the lowest voltage at which a battery is assumed to be fully discharged, the installer should also consider these parameters for determining the appropriate capacity of the battery [22].

5.3.3 Sizing Considerations

Appropriate battery sizing is necessary to fully utilize the benefits of a storage system. A wrong battery sizing can result in irreversible battery damage due to over-discharge, inadequate voltages to the load, and insufficient backup times. The autonomy and the characteristics of the load define the required battery capacity for a PV application. The following parameters are required to calculate the required capacity of a battery in a particular PV system.

5.3.3.1 Selecting Appropriate Voltage
If a DC/DC converter is not incorporated in the system, the appropriate voltage is specified by the nominal voltage of the load (and PV array). This determines how many cells or units need to be connected in series.

5.3.3.2 Defining Maximum DoD
For each battery type, the maximum DoD is defined based on the operating mode. For a LA battery, the maximum DoD is typically set at 80%. The maximum daily DoD can be specified arbitrarily (for example, a value of 20–30% is usual), or it can be calculated using the known daily cycle, the battery's cycle life, and the needed lifespan (if the limiting factor is cycling). A maximum DoD must be established for seasonal storage (if needed). If a LA battery will not be fully charged for several weeks, it is advised not to discharge the battery to more than roughly 30% DoD.

5.3.3.3 Additional Considerations
To ensure choosing an appropriately sized battery, the following points should be considered during battery sizing.

Determination of Duty Cycle
According to IEEE Std 485-2020 [24], duty cycle is "the sequence of loads a battery is expected to supply for specific time periods." The duty cycles of a battery vary with the application. Therefore, it has to be determined at the very beginning of battery sizing. The duty cycle of a battery depends on several factors such as cycle duration, charge and discharge rates, depth of discharge, and length of time in the standby mode. At a specific instant, the knowledge of these parameters allows us to estimate the condition of the battery. The battery duty cycle is measured in Amperes for typical systems, and in Watts for microgrids [25].

Calculation of the Required Overall Energy
The total required energy of the battery may be estimated by integrating the delivered power. The energy capacity also depends on the energy transformation efficiency of the BESS (η_{BESS}). However, it is preferable to incorporate a capacity margin to account for sudden increases in load consumption and system losses. So, 10% is added to the total energy calculated using Eq. 5.6 [25].

$$E_{required\_overall} \, Wh = \frac{110\%}{\eta_{BESS}} \int_0^{24} (P_{demand} - P_{supply}) \, dt, \qquad (5.6)$$

where $P_{demand} > P_{supply}$.

Battery Stack Sizing

A cell is the smallest, most compact form a battery can have, with a voltage ranging from 1.2 to 4.2 V. Several cells are stacked in series or parallel to form a battery module to produce the needed capacity and operating voltage in a battery stack. To get a higher voltage, the cells are connected in series, whereas to get a higher current, the cells are connected in parallel [26]. The output voltage of a module is given by Eq. 5.7.

$$Terminal\ voltage = single\ cell\ voltage \times number\ of\ series\ cells. \qquad (5.7)$$

A string of cells in series has the same Ah capacity as one individual cell, but the voltage of the string is the sum of the individual cell voltages. If the required Ah capacity exceeds the capacity of the available battery module, several similar strings should be parallelly connected to increase the capacity. As a result, the battery module's total Ah capacity is given by Eq. 5.8.

$$Ah\ capacity = Ah\ capacity\ of\ single\ string \times number\ of\ parallel\ strings. \qquad (5.8)$$

After that, a battery pack is put together by connecting modules in series or parallel. Finally, several series- or parallel-connected modules make up a stack.

Estimation of Battery Life Cycle

The battery capacity reduces with time due to continuous cycling, which is referred to as battery aging. As per the United States Advanced Battery Consortium (USABC), a battery should be replaced if its capacity goes below 80% of its nominal value. The storage technology and operating circumstances influence the life cycle of any battery. The relationship between the life cycle and DoD is expressed as Eq. 5.9 [25].

$$N_{cycle} = \alpha \times \Delta DoD^\beta, \qquad (5.9)$$

where α is a temperature-dependent function of the life cycle, and β is a coefficient of life cycle dependent on the DoD; both are battery-specific parameters [27, 28]. Here, ΔDoD is the change of the DoD or the linear cycle depth, which is in the range of 0–1. Simply put, when a battery undergoes a complete charge–discharge cycle, ΔDoD = 1, and if the battery is 50% discharged, and then charged to full capacity, ΔDoD = 0.5 [29]. After the battery has reached the end of its useful life, it may be replaced immediately. The preferable option is to consider the necessary operational period when sizing the battery.

Days of Autonomy

The days of autonomy are the number of days that a battery bank has to provide the power demand of a particular load without any other power source. Since solar energy has an intermittency issue, consumers of solar PV systems need to prepare for autonomy days beforehand. The number of days of autonomy is determined by a variety of factors. In a certain place, based on weather and climate data and solar radiation data, the maximum number of consecutive overcast days is predicted. Another fact that needs to be considered is that critical load applications, such as system applications, require greater autonomy than non-critical load applications [22].

Capacity Calculation

The capacity of a BESS depends upon multiple factors. A battery is sized according to these aspects. In the case of seasonal storage, the "seasonal Ah" requirement means the quantity of storage necessary to compensate for a seasonal deficit in PV array production. It can be expressed as Eq. 5.10.

$$C1 = \frac{Seasonal\ Ah}{Maximum\ seasonal\ DoD}. \tag{5.10}$$

For autonomy storage, a number of days should be specified for which the battery needs to supply the load in the event of an emergency. If there is a chance for the site to come across sub-zero conditions (sub-zero temperatures are below $0\ °C$ or, in the USA, below $0\ °F$), the normal DoD limit might be modified to avoid freezing the battery. Considering all these factors, the capacity can be represented as Eq. 5.11.

$$C2 = \frac{Average\ daily\ load, ssh \times Days\ of\ autonomy + seasonal\ Ah}{Maximum\ DoD\ (adjusted\ to\ prevent\ freezing\ if\ necessary)}. \tag{5.11}$$

For the desired battery capacity to meet the daily cycling demand, it is essential to know the duration that the load has to be energized (24 h continuously for a day, only during the day, or only during the night). For continuous loads, it is expected that the battery will go through 6–8 h of charging in a day. So, the daily discharge Ah would be calculated by multiplying the daily load in Ah by a factor of (16/24) to (18/24) following Eq. 5.12.

$$C3 = \frac{Daily\ Ah\ discharged}{Maximum\ daily\ DoD}. \tag{5.12}$$

And lastly, the maximum charging rate should not exceed the limit specified for the battery. The maximum charging rate is the rate at which the system can charge, and it is unique for each battery. To compute it, we need to know how much current the PV array can produce under the highest sunshine. Equation 5.13 is followed to find the battery capacity in such a case.

Table 5.3 Ambient temperature multiplier for battery [30]

Ambient temperature, °F (°C)	Multiplier
120 (48.9)	0.86
110 (43.3)	0.88
100 (37.8)	0.91
90 (32.2)	0.94
80 (26.7)	1.00
70 (21.2)	1.04
60 (15.6)	1.11
50 (10.0)	1.19
40 (4.4)	1.30
30 (−1.1)	1.40
20 (−6.7)	1.59

$$C4 = \frac{Battery\ maximum\ C\ rate,\ hours}{Maximum\ array\ current,\ A}. \tag{5.13}$$

C1, C2, C3, and C4 are all expressions for the battery capacity. The highest of these four values is finalized as the desired battery capacity [22].

Temperature Effect

Extremes in temperatures can damage batteries. A cold battery cannot produce as much energy as a warm one. Even though a heated battery may deliver more than its stated capacity, operating at high temperatures reduces battery life. Winter ambient temperature multiplier should be considered while sizing the battery to ensure that the battery capacity is maintained over the winter season.

From Table 5.3, it can be seen that there is a multiplier for a specific temperature where the batteries will be stored. Since colder temperatures will impact battery capacity, this multiplier should be considered when the room temperature is less than 26 °C.

5.3.4   Battery Sizing Example

To compensate for the number of autonomy days, the battery needs to satisfy the steady energy demand of solar PV loads. Below is an example of sizing a battery considering autonomy days.

Example 5.1 (Battery Sizing) In this example, the battery sizing process is discussed, taking autonomy days into account.

Suppose a system requires 900 Wh energy at 24 V, and it is using a battery bank that has a voltage of 12 V. First, to calculate initial capacity, energy usage is divided by battery bank voltage. So:

$$Initial\ capacity\ the\ battery\ needs\ to\ provide = \frac{900}{24} = 37.5\ Ah.$$

Next, the capacity is multiplied by the days of autonomy to get the desired battery capacity. Typically, the days of autonomy are around 2–3 days. So, the required capacity becomes

$$Required\ battery\ capacity = 3 \times 37.5 = 112.5\ Ah.$$

Next, the DoD is considered, which is generally around 50%. Therefore, the required capacity for the DoD can be calculated using

$$Capacity_{DoD} = \frac{Capacity_{required}}{DoD} = \frac{112.5}{0.5} = 225\ Ah.$$

If this solar battery is in a room where the temperature is below 26 °C, a multiplier would be required. This specific multiplier then will need to be multiplied with the capacity to compute the final amount. Suppose the ambient temperature is 70 °F or 21.2 °C (according to Table 5.3, the capacity has to be multiplied by 1.04), the required overall capacity would be

$$Final\ capacity = 225 \times 1.04 = 234\ Ah.$$

Considering these parameters, the required energy by the system for 3 days of autonomy would be

$$234\ Ah \times 24\ V = 5.62\ kWh.$$

Suppose the battery used in this example is a Chargex 12 V 60 Ah Li-ion battery. Before purchasing the batteries, it is essential to determine the number of batteries needed to meet the required energy. Any combination of series and parallel connection of batteries can be used to obtain this energy. To find out the number of batteries connected in parallel, which will serve the required capacity, the following method is used:

$$No.\ of\ batteries\ in\ parallel = \frac{Required\ total\ battery\ capacity}{Battery\ Ah\ rating}$$
$$= \frac{234}{60} = 3.9 \approx 4.$$

If the result is not an integer, it is rounded up to the next integer number. To calculate the number of batteries connected in series, simply divide the system voltage by battery voltage. As the system's nominal voltage is given as 24 V, the number of batteries connected in series is

$$No.\ of\ batteries\ in\ series = \frac{24}{12} = 2.$$

So the total number of batteries would be

$$Total\ no.\ of\ batteries = 4 \times 2 = 8.$$

Here, 2 branches of batteries are connected in series, and 4 batteries are connected in parallel in each of these branches. Batteries connected in series add up the voltage, while parallel-connected batteries add to the Ah.

To confirm the calculation is correct, we can calculate the total energy provided by the total number of batteries. Total provided Ah is $4 \times 60\,Ah = 240\,Ah$, which is slightly higher than the required capacity of 234 Ah. Total voltage of two series-connected bank is $12\,V + 12\,V = 24\,V$. Therefore, the total energy provided by the battery banks would be

$$240\,Ah \times 24\,V = 5.76\,kWh,$$

which is slightly higher than the required energy demand of 5.62 kWh during the autonomy days. The entire calculation is carried out according to the battery sizing worksheet in Table 5.4. The connection of the batteries has been illustrated in Fig. 5.19.

5.3.5   Battery Sizing Worksheet

This worksheet is used to determine the size of the battery bank required to supply power during the days of autonomy. To begin the calculation, we need to determine the total Ah utilized by both AC and DC loads each day. Table 5.4 provides a generic battery sizing worksheet.

5.3.6   Battery Maintenance

Maintenance is critically essential to get the most life out of a battery. Most types of batteries require maintenance on a regular basis. For example, the following maintenance steps could be undertaken to ensure the healthy operation and an extended lifetime of the battery bank. Some common ways of battery maintenance are delineated below [22]:

- **Water insertion:** Usually, sealed batteries do not require water insertion. Nevertheless, they should be examined regularly, and their terminals should be cleansed. The addition of distilled water is required for open batteries to prevent unwanted chemical reactions with the electrons present in normal water.

Table 5.4 Battery sizing worksheet

Step no.	Step	Unit
1	Enter the daily Ah requirement	Ah/day
2	Enter the days of autonomy required to support the system. (Typically 3–5 days)	Days
3	Multiply the Ah requirement by the days of autonomy. This is the primarily required capacity	Ah
4	Enter the DoD for the battery. The DoD can be different for different battery chemistry. This is important to avoid over-draining the battery	%
5	Divide Ah by DoD to get the secondary capacity	Ah
6	Select the multiplier from Table 5.3 according to the average wintertime ambient temperature the battery will experience. From a hot battery, more than the rated capacity can be extracted, but this shortens the battery life. So, it is preferred to keep batteries at room temperature	–
7	Multiply the multiplier by the secondary capacity to negate the effects of cold weather. This is the total required capacity	Ah
8	Enter the Ah capacity of the chosen battery	Ah
9	Divide the required Ah capacity by the battery Ah rating and round off to the next integer to get the number of batteries required in parallel	–
10	Divide the nominal system voltage (12 V, 24 V, or 48 V) by the battery voltage. Round off to the next integer to get the number of batteries in series	–
11	The product of the series and parallel number of batteries is the required total number of batteries	–

- **Periodic checking:** To recognize a faulty block or cell, a periodic maintenance procedure should be undertaken. It can be carried out by measuring specific gravity, checking the voltage, and checking the temperature of the case.
- **Replacement:** If any faulty cells are detected, they must be replaced. New cells should not be mixed with old ones in one series string. If any cell or block needs to be replaced, it is preferable to replace all parts in a single string.
- **Running under ideal conditions:** Despite all the maintenance, it will not improve the condition of the battery if it cannot run under ideal circumstances. It is best to fully charge the battery as practically possible through charge or voltage regulation. A certain amount of overcharging is common for batteries, and in some cases, it is recommended. But an excessive amount will overall harm the battery.
- **Ventilation:** All batteries produce a good volume of heat during overcharge. That is why ventilation is necessary. In open or vented batteries, the absence of proper ventilation could lead to an explosion hazard. Sealed batteries also need some ventilation for cooling purposes, but not so much for removing the small hydrogen gas they produce.

Unlike conventional batteries, rechargeable batteries used in solar PV systems must function under different conditions. Due to the intermittency of solar energy, these batteries undergo irregular charging and discharging. Cycling factors, operat-

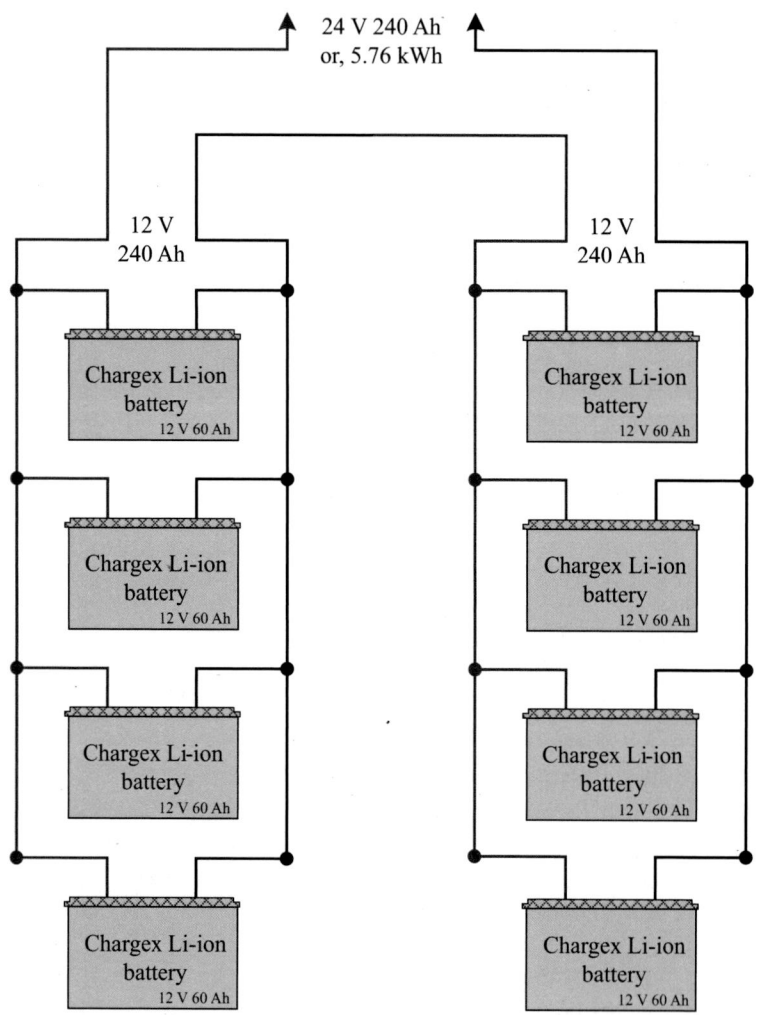

Fig. 5.19 The connection of the batteries according to the given battery sizing example

ing temperature, regional weather, and other factors such as the battery's internal corrosion are important in determining the battery's lifetime in a PV system.

5.3.7 Float Charger

The voltage at which a battery is kept after it has been completely charged in order to compensate for self-discharge is called float voltage. Float voltage depends on the construction of the battery, the chemicals inside it, and the ambient temperature. A float charger will charge a battery at the same pace that it will self-discharge,

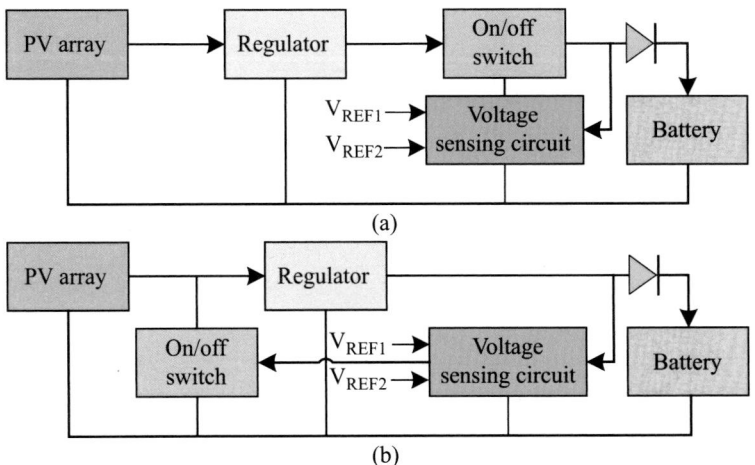

Fig. 5.20   (**a**) A switched series float charger and (**b**) a switched shunt float charger [31]

ensuring that the battery remains fully charged. It has inner circuitry that prevents the battery from overcharging.

Float chargers can be connected in the circuit in two ways: series and shunt. These configurations are illustrated in Fig. 5.20. Typically, float charging is employed in emergency and backup applications where the battery is only discharged occasionally. When the battery voltage hits a preset reference level (V_{REF1}), such as full charge, this charger detects it and turns off the power to the battery. It switches the power to the battery back on when the battery drains to a second preset level (V_{REF2}). The inner voltage sensor circuitry of the float charger juxtaposes the battery voltage with V_{REF1} and V_{REF2} and activates or deactivates the shunt switch correspondingly. Another form of float charger contains the on–off switch in series with the load and source in which the power is set by a regulator. The scheme of the working principle of the series switched float charger is illustrated in Fig. 5.21.

5.4   Regulators or Charge Controllers

Batteries are sensitive electrochemical energy storage devices that must operate under specific operating conditions provided by the manufacturers. For this reason, charge controllers or electronic regulators are essential devices in all battery-based PV systems. Installed between the PV module and the load, charge controllers prevent batteries from overcharging during the daytime when the demand is low compared to generated power. They also prevent the deep depletion of the battery when the load demand is high and draws a high current for the long term. So,

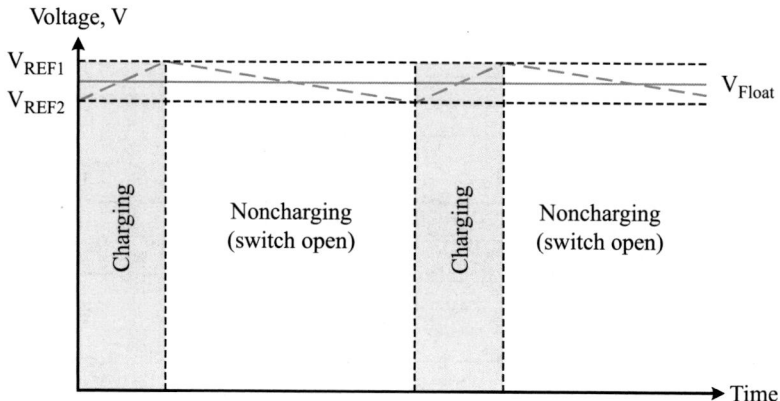

Fig. 5.21 The graphical overview of the working of a switched series float charger [31]

charge controllers are responsible for both charging and discharging processes of the battery.

Charge controllers monitor the battery voltage or temperature to determine the current SoC. Based on the SoC, the pre-programmed controller decreases, increases, connects, or disconnects the current provided to the battery or by the battery. The charge controller is responsible for controlling both the charging and discharging processes. On one side, the controller plays the role of a discharge regulator by opening the electric circuit between the battery and the load to avoid the negative impacts of over-discharge, which include loss of active mass, severe degradation in capacity, and sometimes salvaged batteries. On the other side, it works as a charge regulator by controlling a switch that opens or closes the circuit connecting the solar PV array and the battery system, when the battery is fully charged or needs some charging current, respectively. It prevents the gasification process in VRLA batteries or case damage in other lead-acid batteries. The negative consequences of the absence of a charge controller are much worse in lithium-ion batteries, which have a very high energy density and low overcharge tolerance that might lead to extreme temperature rise. The rise in temperature might end up with a thermal runaway followed by an explosion.

These regulators usually come with specific voltages and a current rating, such as 12 V, 24 V, or 48 V and 15 A, 20 A, 30 A, 50 A, or 80 A. Charge controllers can operate at a constant voltage, i.e., equal to the battery's voltage. For such cases, pulse width modulation (PWM) controllers are used. Moreover, these controllers can operate at a higher or a wide range of voltages. In such cases, algorithm-controlled hardware is embedded with the controller circuit called Maximum Power Point Tracker (MPPT).

Solar charge controllers are designed to operate at a specific range of PV module voltages. For example, if a PV system is formed using a parallel connection of several 18 V solar modules, a 12 V or 24 V controller can be connected with the system. If two 60-cell modules are connected in series, the V_{mpp} of the system can be found from the datasheet or the following equation:

$$V_{mpp} = N_{module} \times N_{cell} \times V_{cell}$$
$$= 2 \times 60 \times 0.5\ V \qquad (5.14)$$
$$= 60\ V,$$

where

N_{module} = the number of modules in series,
N_{cell} = the number of cells in each module, and
V_{cell} = cell voltage.

This system is usually connected with a 48 V battery system, where four 12 V batteries are connected in series. Therefore, a 48 V charge controller is ideal in this case. As we know, the current flow path is from a higher potential place to a lower potential place. So, to maintain the current flow from the PV modules to the load (battery), the voltage of the PV array must be kept higher than the battery or controller voltage.

5.4.1 Charge Regulation Set Points

The charge regulation set points determine the controller's behavior in the charging phase to ensure the recommended operating conditions. Some controllers can come with an ability to set these points by the designer/installer, and some come with built-in fixed points that are universal and agreed by the battery manufacturers. The charge regulation set points come with a pair of points as discussed below:

5.4.1.1 Voltage Regulation (VR)
The VR set point is the highest voltage of the battery, above which the controller stops the current flow from the PV array to the battery. The value of this point varies based on the battery chemistry and the surrounding temperature at which the battery is operating. For instance, lead-acid batteries in average room temperatures have a VR value of 2.2–2.3 V per cell, which equals 13.2–13.8 V for a 12 V, 6-cell battery module.

5.4.1.2 Voltage Regulation Hysteresis (VRH)
The VRH is the difference between the VR and the voltage at which the full current of the PV array is reconnected. The charge controller supplies current to the battery to ensure that the battery is fully charged. The difference between VRH and VR values determines the complexity of the controller. The lower the difference, the lower is the charge interruption time, and so the design complexity of the charge controller will be higher. The lower the VRH value, the more is the noise and

harmonics. Thus, a higher switching rate will be required for the switching element in the controller.

5.4.2   Discharge Regulation Set Points

Similar to the charge regulation set points, discharge regulation set points are the parameters at which the controller starts taking actions to protect the battery from the over-discharge phase. The discharge regulation set points also come with a pair of points as discussed below:

5.4.2.1 Low Voltage Disconnect (LVD)

The LVD is the lowest possible voltage at which the battery is disconnected from the load to prevent over-discharging. In other words, it is the set point at which the controller opens a switch to protect the battery from being depleted under the desired value. The LVD value varies based on the system designer's vision about the allowable depth of discharge and how the energy storage system is oversized or critically sized. The higher the LVD, the lower the allowable depth of discharge and the higher the cycle life of the batteries. For a 12 V, 6-cell lead-acid battery, the LVD can range from 1.8 V to 1.93 V per cell, which means the LVD at the module level (combining 6 cells) ranges from 10.8 V to 11.6 V.

An example of an LVD module is the CZH-LABS Low Voltage Disconnect Module [32], which is rated at 12 V and 30 A. Its operating voltage is 6 V to 18 V DC. It has four terminals: two for the battery and two for the load. It has both LVD and LVDH settings and operates using microcontroller and MOSFET switches.

5.4.2.2 Low Voltage Disconnect Hysteresis (LVDH)

The LVDH is the difference between the LVD and the voltage at which the battery is reconnected to the load. Once the battery (load) is disconnected from reaching the LVD value, the battery's voltage will have an immediate voltage rise or recovery that does not necessarily refer to a good SoC. However, it will fall again if the load is reconnected. For this reason, it is crucial to keep the LVDH high to avoid the misleading recovery value of voltage rise and to protect the load from being on and off very frequently.

Although charge controllers have a specific objective to protect the battery and avoid any harmful operation circumstances, this objective might be achieved using very simple technology. Using electromechanical relays or power electronic devices is preferable, which works in the range of kHz and MHz.

5.4.3   Temperature Compensation

A solar module operates by absorbing sunlight and transforming it into electricity. But not all of the absorbed sunlight is turned into electricity; rather a part of this collected light is lost as heat. Thus, the overall temperature of the solar module rises.

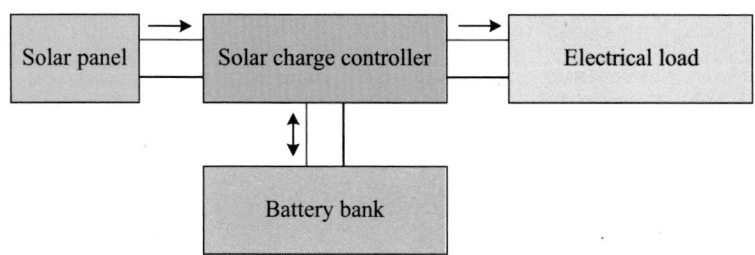

Fig. 5.22 Solar charge controller in a PV system

This increase in temperature can drastically affect the efficiency of the solar module. All solar cells have a temperature coefficient in their datasheet, typically measured in %/°C or %/K. The temperature coefficient of a solar module indicates how much its efficiency will drop or rise as the module's temperature rises or falls from the standard temperature of 25 °C. Typically, the temperature coefficient of a solar PV ranges from -0.35% to -0.5%/°C. Therefore, when a module's temperature increases up to 35°C on a hot day, its efficiency drops by $0.5 \times 10 = 5\%$. Again, if the temperature falls to 15 °C, then the efficiency of the PV module increases by $0.5 \times 10 = 5\%$. The operating temperature of a PV module ranges from -40 °C to $+85$ °C. The ambient temperature affects battery performance as well. The efficiency of the battery can drop due to cold weather as it will get undercharged. So, battery charging voltages should be corrected based on temperature. This is referred to as temperature compensation, where a battery is charged to a lower voltage in a warm temperature and greater voltage in a cold condition.

A charge controller or a battery regulator helps prevent overcharging of a battery by limiting the flow of electric current rate to and from the battery. The connection of a charge controller in a solar PV system is depicted in Fig. 5.22. The battery can be charged effectively in both warm and cold conditions by regulating the charge voltages for the respective temperature.

In today's solar power systems, two types of charge controllers or regulators are used: maximum power point tracking (MPPT) and pulse width modulation (PWM). These regulators monitor the temperature of the battery to prevent overheating. Depending on the battery's maximum capacity, these regulators also adjust the charging rate. Since MPPT operates above the battery voltage, it is more efficient in cold temperature and outperforms PWM as it can provide a "boost." But PWM works better in warm temperatures as it operates at battery voltage [33].

Automatic temperature compensation is a widespread solution in most PV charge controllers and inverters in today's world. One of the batteries in the bank has a remote temperature sensor (RTS) attached to it. The charger utilizes the data received from the RTS to determine the temperature of the battery and automatically adjust the target charge voltage as per the requirement, as illustrated in Fig. 5.23.

The temperature adjustment formula that is most commonly used is: -0.005 V per °C per 2 V cell [34]. The other values deployed are -3 mV/°C/cell or

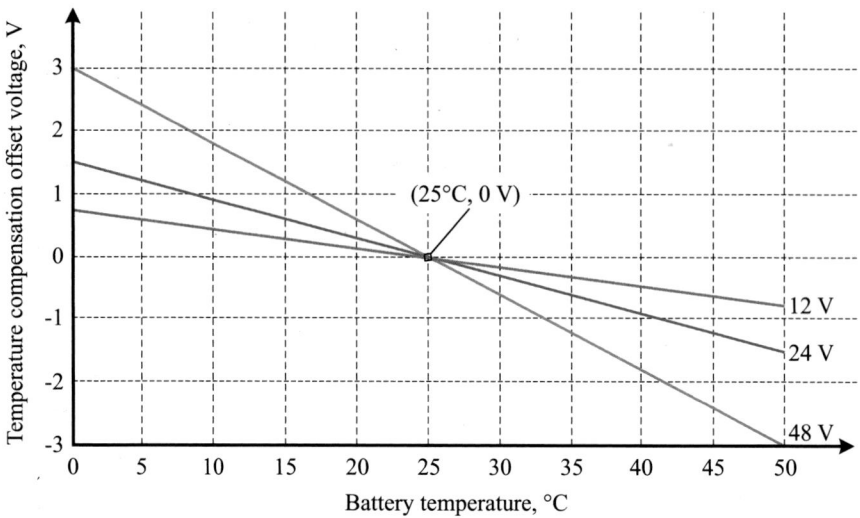

Fig. 5.23 The slope of typical temperature compensation offsets for lead-acid cells varies based on nominal battery bank voltage [34]

-4 mV/°C/cell [35]. In order to understand and apply this formula, three parameters specific to a given system should be known—battery temperature, reference temperature, and nominal battery voltage. Most chargers have a standard reference temperature of 25 °C, at which the batteries usually have the best combination of energy storage capacity and life cycles. Depending on the system capability, the charger raises the target voltage if the RTS result indicates the battery temperature is below 25 °C and decreases the target voltage if the RTS value indicates the battery temperature is above 25 °C. The number of cells that make the battery bank determines the nominal battery voltage. Virtually, almost all lead-acid battery cells have a nominal voltage of 2 V_{DC} per cell. So, depending on the number of cells, the nominal voltages of strings of batteries can be determined as follows:

$$6\ cells = 12\ V_{DC},$$

$$12\ cells = 24\ V_{DC},$$

$$24\ cells = 48\ V_{DC}.$$

In the example that follows, two different operating temperatures are considered to calculate the temperature compensated voltage of the charge controller.

Example 5.2 (Temperature Compensation in Charge Controller) In this example, how the voltage of a charge controller is compensated for a particular temperature is discussed.

Let us consider a 12-cell, 24 V battery bank operating at 15 °C. So, the temperature deviation in this case is $15 - 25 = -10$ °C. The temperature compensation voltage for this battery would be

$$-0.005\ V \times -10 \times 12 = 0.6\ V.$$

If the nominal battery voltage recommended by the manufacturer is 27.6 V, then the temperature compensated charge voltage for that battery bank at 15 °C would be

$$27.6\ V + 0.6\ V = 28.2\ V.$$

Again, at an operating temperature of 35 °C, the temperature deviation is $35 - 25 = 10$ °C. So, the temperature compensation voltage will be

$$-0.005\ V \times 10 \times 12 = -0.6\ V.$$

In this case, the temperature compensated charge voltage for that battery bank at 35 °C would be

$$27.6\ V - 0.6\ V = 27\ V.$$

5.4.4 Shunt Charge Controller

Shunt charge controllers prevent the batteries from being overcharged by shunting the solar array using a parallel controlled switch. As known in the electrical design field, a short-circuit current is not a preferable current. Electrical engineers usually design every system to be protected from such a current. It is essential to know that the use of shunt charge controllers is limited to a very small solar PV system where the current capacity does not exceed 20 A. Moreover, the short-circuit current passing through the cables must be dissipated in the controller through a heat sink. So, it is critical to ensure sufficient ventilation to the controller. The controller is also connected to a switch on the load side to open the circuit when the battery voltage is too low.

Figure 5.24 shows that the battery is not shorted with the shunt or parallel switch. However, they are connected by a diode known as a blocking diode. This diode mainly plays two roles. First, it prevents the battery from being shorted when the switch is closed. Second, since a diode blocks the reverse flow of current through it, it prevents the solar array from being an extra load. Shunt charge controllers follow the on–off system. Current flows from the PV module to the battery until it reaches VR, which means until the battery is fully charged. When it reaches VR, switch 1 closes to provide an alternative route for the current to pass through the module (short circuit) and not the battery until the voltage reaches the LVD state. When the battery voltage reaches LVD, switch 2 will open to prevent the battery from over-discharging or over-depletion.

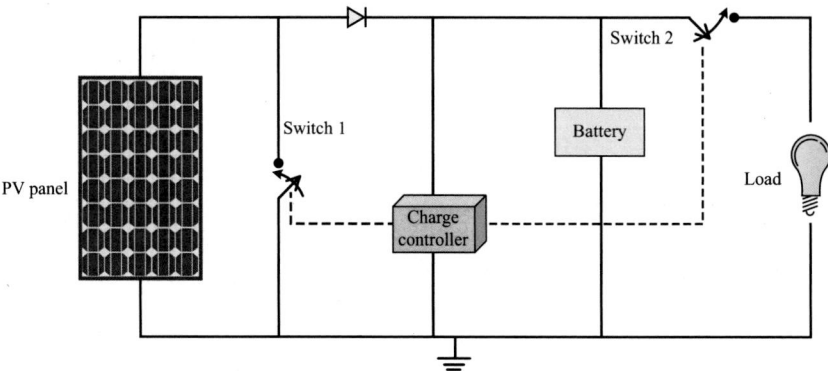

Fig. 5.24 Block diagram of a shunt charge controller

5.4.5 Series Charge Controller

In contrast to the shunt controllers, series controllers control the charging process using a series switch between the battery and the PV array. This technology is superior to the shunt charge controllers due to its practicality in large-scale arrays. In addition, the series charge controller has no current limitation, and the controller can easily handle the open-circuit voltage.

The block diagram of a series charge controller is illustrated in Fig. 5.25. Observe that there is no diode in this controller, in contrary to the shunt charge controller. Similar to the shunt charge controllers, series charge controllers follow the on–off system. However, switch 1 opens the circuit in this configuration, which blocks the current from flowing from the PV array to the battery when it reaches VR. When the battery reaches LVD state, switch 1 closes the circuit, allowing the battery to get charged. At this point, switch 2 is opened to avoid over-discharging. When the LVDH value is reached, switch 2 is closed to restore power supply.

5.4.6 Pulse Width Modulation Controller

Pulse width modulation (PWM) controllers are similar to series charge controllers, but in this case, a transistor is used. PWM controllers control the flow of current based on the width of the signal passed to the transistors, as shown in Fig. 5.26. Here, the overall voltage is kept constant no matter what amount of modulation is done.

PWM controllers operate following an on–off state. The signal passed to the transistor is illustrated in Fig. 5.27. The amount of current flowing through the controller depends on how long the controller is switched on. In Fig. 5.27, the rectangles represent the pulse, turning on-state and turning off-state. For example, for 70% modulation, the controller is turned on for 70% of the time and turned off

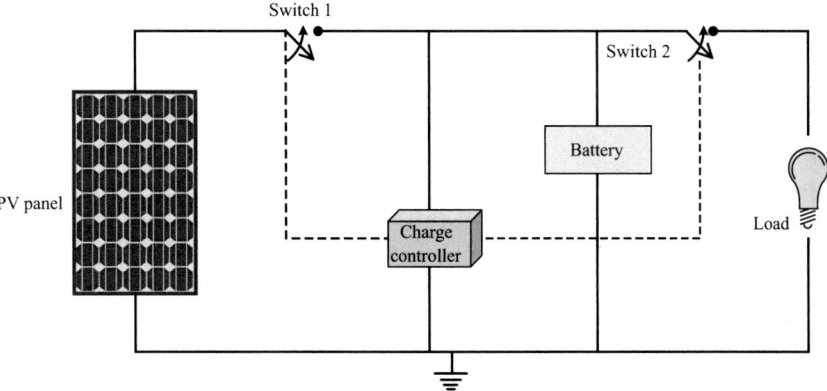

Fig. 5.25 Block diagram of a series charge controller

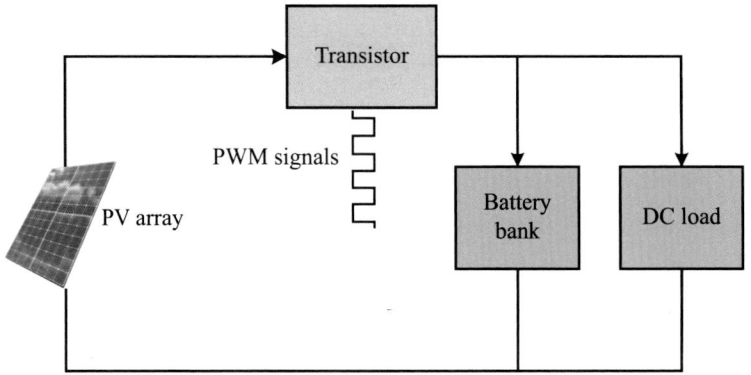

Fig. 5.26 The PWM controller connected to a solar PV system

for 30% of the time. This continues over and over again. It means that we are only letting the controller pass 70% of the available power. This on and off phenomenon occurs rapidly, about thousands of times per second.

This is a precise way to control the flow of power from the PV array to the battery bank and the loads. The drawback of the PWM charge controller is that it produces a lot of noise, such as radio frequency or electrical noise. But some controllers come with filters embedded in them, which reduces a lot of noise. Usually, for simpler circuits, PWM controllers are used.

5.4.7 Maximum Power Point Tracking Controller

Maximum power point tracking (MPPT) controllers are power electronics devices embedded in most new inverters and charge controllers. Since solar modules

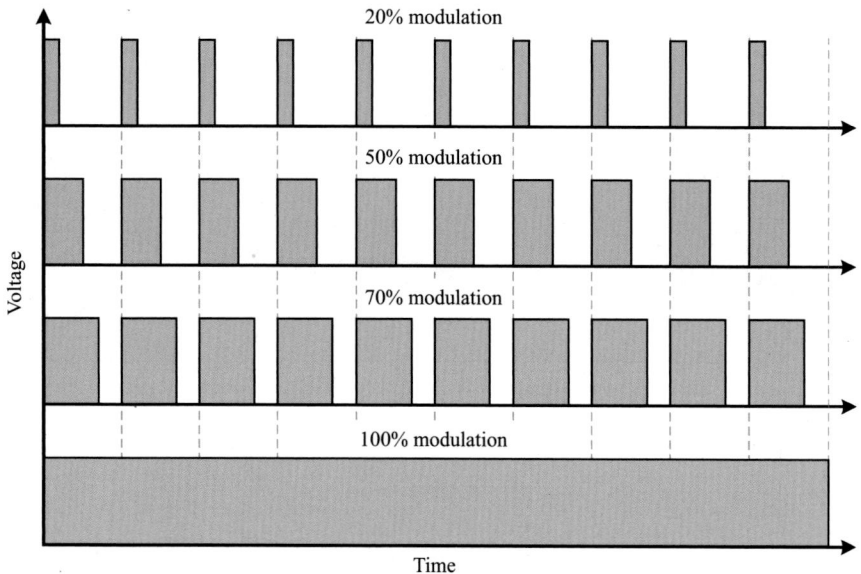

Fig. 5.27 Modulating signals that are input to the PWM controller

are voltage-controlled current sources, it is important to have a technique that guarantees that the solar array works in an appropriate voltage, which achieves the maximum current out of the array. This voltage, called V_{mpp}, and its matched current called I_{mpp} are the planar coordinates of the maximum power point (MPP) in the I-V curve, as discussed in Chap. 3.

Calculating and tracking this voltage is an easy process if the source and the load have constant characterizations. However, solar modules have different I-V curves based on different conditions, such as irradiance, temperatures, and shading conditions. In addition, the load can easily be varied based on the electrical appliances' usage or the grid demand. Thus, complex algorithms were developed to track V_{mpp} accurately in real time to ensure that the PV array operates at the maximum power point. A basic representation of using an MPPT charge controller in a PV system is demonstrated in Fig. 5.28.

5.4.7.1 Commonly Used Techniques in MPPT
The commonly used techniques in MPPT controllers can be primarily classified as indirect and direct. In this section, two indirect methods and two direct methods are explained.

Indirect Methods
- **Fixed voltage method:** It is the simplest technique that requires easy periodic assumptions. However, this method is based on modifying the operating voltage value into fixed seasonal values provided by the designer. Besides, it highly

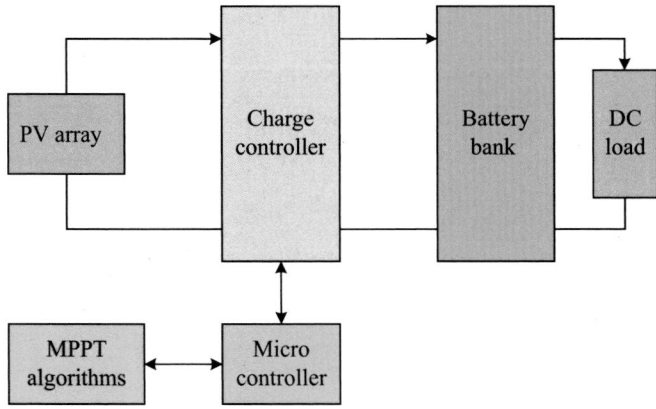

Fig. 5.28 Basic schematic of an MPPT charge controller connected in an off-grid solar PV system

depends on the effect of temperature. Thus, the PV modules work at a higher voltage in the cold season (winter), a relatively low voltage during the hot season (summer), and in between in both fall and spring. This method is not accurate unless the solar array's location has a neglected irradiance fluctuation.

- **Fractional open-circuit voltage:** This method is not based on pre-programmed seasonal values. Here, the open-circuit voltage, V_{oc}, of the module is measured and then multiplied with a constant K to obtain the V_{mpp} as shown in Eq. 5.15.

$$V_{mpp} = K \times V_{oc}. \tag{5.15}$$

Direct Methods

- **Perturb and Observe (Hill climbing method):** The main concept of this algorithm is to provide a deviation or perturbation to the voltage of the PV array and observe the corresponding changes in power. Recall the position of the MPP in the characteristic curve of a PV cell. In a PV cell, if an increase in voltage causes the power to rise, then the voltage is kept on increasing up to the point when the power begins to fall again. Once the output power starts to decrease with the increase of voltage, the controller understands that the MPP has already been reached. Thus, the MPP is detected.
  This method is widely used because it is inexpensive and has a better performance rate than the indirect methods. However, the operating point will not fully meet the MPP but be close to the MPP.
- **Incremental conductance (INC) method:** This method is highly efficient as it uses the INC algorithm. The main function of this algorithm is to find the slope of the P-V curve. Since MPPT charge controllers focus on attaining the MPP, INC algorithms detect the peak value of the P-V curve with the help of the obtained slope. This process is conducted using the basic formula of power. The basic equation of power is

$$P = V \times I. \tag{5.16}$$

Differentiating Eq. 5.16 with respect to voltage V, we get

$$\frac{dP}{dV} = \frac{d(V \times I)}{dV}.$$

We know, at the maximum point of any curve, the slope is always zero. Here, $\frac{dP}{dV}$ represents the slope of the P-V curve. Therefore, at MPP, the slope is zero, i.e.,

$$\frac{dP}{dV} = I \times \frac{dV}{dV} + V \times \frac{dI}{dV} = I + V \times \frac{dI}{dV} = 0.$$

So, at MPP,

$$\frac{\Delta I}{\Delta V} = -\frac{I}{V}.$$

Two other cases can be seen here:

1. If $\frac{\Delta I}{\Delta V} > -\frac{I}{V}$, the operating point of the system is to the left of the MPP.
2. If $\frac{\Delta I}{\Delta V} < -\frac{I}{V}$s, the operating point of the system is to the right of the MPP.

The controller checks $\frac{\Delta I}{\Delta V}$ ratio and compares it to the negative value of the conductance (G $= \frac{I}{V}$) to know the position of the operating point with respect to the MPP. Figure 5.29 illustrates the steps of the INC algorithm in a flow chart.

5.4.8   Comparison between PWM and MPPT Charge Controller

Both PWM and MPPT charge controllers are used in solar PV systems to manage their operation. MPPT is known as the advanced version of PWM as it has the ability to vary the PV array voltage from the battery bank voltage. A comparative analysis of these two types of charge controllers is provided below:

- **Temperature compensation:** Most of the PWM charge controllers do not have temperature sensors in them, which means the system will lack temperature compensation. Thus, the battery will die out sooner as the change in temperature will affect the output and the system. On the other hand, MPPT charge controllers have temperature sensors in them, which allows compensating the system temperature.
- **Voltage limitation:** There is a voltage limitation in the case of PWM. If the system generates more voltage than its rating, the extra energy will be dissipated as heat, whereas, for MPPT, the voltage rating can be increased by keeping the

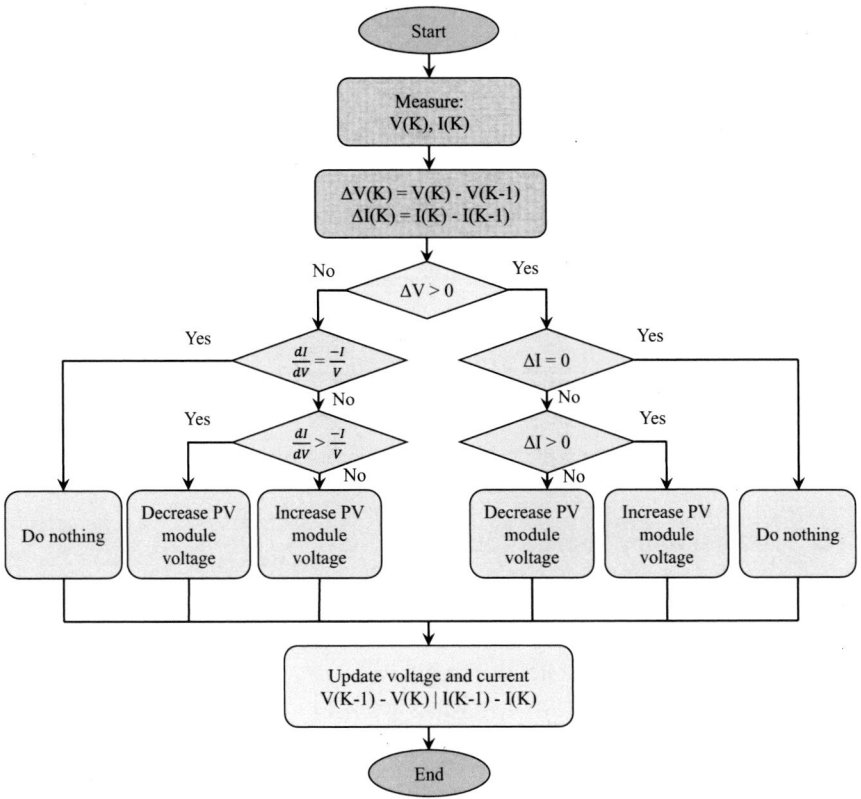

Fig. 5.29 Flowchart of the incremental conductance algorithm

rate of current constant. We take two MPPT charge controllers, for instance. One with 12 V, 20 A, 260 W and the other with 24 V, 20 A, 520 W. We can use 24 V instead of 12 V, where we will have more power rating if we want to use a higher voltage battery.

- **Price:** The price of PWM controllers is cheaper than the price of MPPT controllers.
- **Use of conductors:** The current rating of MPPT (up to 80 A) is higher than PWM ($\leq$60 A) [36]. For this reason, thicker wires are used in the case of MPPT and thinner in the case of PWM.
- **Size of system:** For smaller and simpler systems, PWM charge controllers are preferred. And for larger, utility-scale systems, MPPT charge controllers are preferred, where the current in each inverter can reach multiple kA.

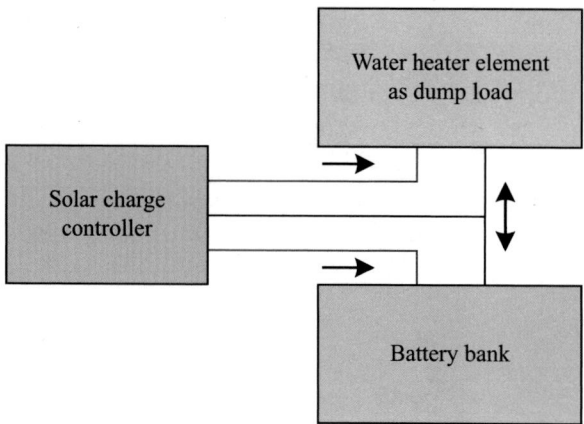

Fig. 5.30 A diversionary charge controller (with the dump load wired to it)

5.4.9 Diversionary Charge Controller

When a battery is completely charged, a diversionary charge controller manages the charging current by directing surplus power to a secondary load. To understand how a diversionary load controller works, it is necessary to know about dump loads first. A dump load is an extra resistive load connected in parallel to the actual load or to the battery to drain the system's surplus power. These loads are solely used to keep deep cycle batteries from overcharging and causing damage. A diversion charge controller controls the dump load. Figure 5.30 shows how a diversionary charge controller is connected to a dump load. The controller only operates to divert the extra current away when the batteries reach the desired voltage. For the rest of the time, it does not take any action.

5.4.10 Charge Controller for Hybrid PV Systems

Charge controllers for hybrid PV systems are required to simultaneously handle several power sources. A hybrid system combines different energy sources to ensure continuous power. A hybrid solar PV system is a grid-tied PV system that has a battery storage system for storing backup power for an unexpected grid power outage. This system allows the battery to be charged by either grid power or solar power.

The loads, in general, are divided into two categories—critical load and non-critical load. Critical loads are emergency loads that require an uninterrupted power supply, such as the machinery used in hospitals and industrial processes. Even in the case of a grid power outage, such loads should be fed by a backup power supply to ensure continuity of power. These loads are vulnerable to unusual or irregular electrical power supplies to communication centers, equipment reliability (especially in hospitals), personnel safety or security, and military base. On the other hand, non-critical loads are the type of loads that can be switched off during a power failure to save power from the backup storage, for example, printers and desk fans.

In a solar PV system, both AC and DC power systems are incorporated. When a battery is installed, the output power of the solar array can flow through two paths—from the module to the DC loads as DC or from the module to the battery to the inverter to the loads as AC. So, a major difference between AC- and DC-coupled systems lies in the path taken by the current produced by solar PV modules and the nature of the current flowing.

DC-coupled System

A DC-coupled system is one in which the form of current flowing through most of the circuit is DC. In such a system, the PV array sends power to a charge controller, which feeds a battery bank. Next, a battery-based inverter converts the DC output from the battery into AC and sends it to the grid after synchronization or directly powers the connected loads. When the utility grid goes down or is disconnected for purposes such as lower power quality, safety, and high energy prices, the battery-based inverter disconnects the entire system from the grid. The disconnection is necessary for the safety of the grid workers; if not disconnected, the power generated from the solar array would continue to flow toward the grid, making it risky to perform maintenance work on the grid. While the system is isolated from the grid, the battery supplies power to the loads through the battery-based inverter.

AC-coupled System

An AC-coupled system is one in which the form of current flowing through most of the circuit is AC. In such a system, the DC power generated from the solar array flows through a solar inverter that transforms the DC power to AC power, which is then either supplied to the connected loads or sent as input to a multi-mode or battery-based inverter. The inverter can synchronize with the utility grid and feed AC power to the grid, or provide DC power to charge the battery bank to store energy. When the grid is down, the battery-based inverter powers the loads from the battery. Since the output of the solar inverter is also connected to the switchboard, it senses the output of the battery-based inverter and assumes that the grid is still operational. So, as long as sunlight is available, the solar inverter keeps sending solar power to the battery-based inverter to feed the battery as well as the connected loads. During the night or on a cloudy day, the battery uses this stored power to run the loads.

Figure 5.31 juxtaposes the DC-coupled system and the AC-coupled system to help compare the two systems and the flow of current through the various components.

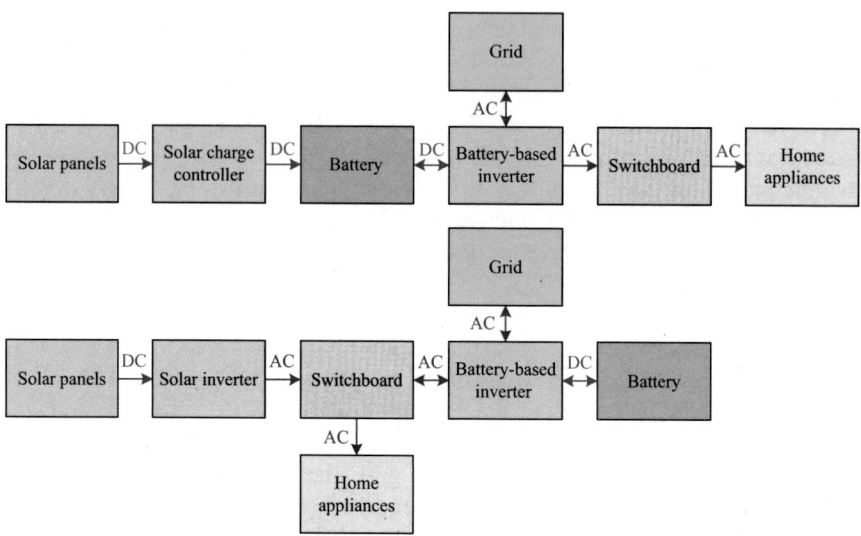

Fig. 5.31 Schematic of a DC-coupled system (top) and an AC-coupled system (bottom)

5.4.11 Charge Controller Sizing

Solar charge controllers are installed in between solar PV modules and batteries and are sized based on the PV module's voltage and ampacity. They help to enhance the efficiency and life cycle of the entire solar PV system. The controller needs to be large enough to manage the power produced by the solar PV modules. Solar PV charge controller ratings may vary from 6 V to 600 V and from 1 A to 80 A [37]. However, the typical voltage ratings are 12 V, 24 V, and 48 V, and the typical current ratings are 10 A, 20 A, and 40 A. An example of sizing a charge controller is given below.

Example 5.3 (Charge Controller Sizing) This example discusses how a charge controller is sized for a particular solar PV system. Let us assume that a solar PV system has a rated voltage of 12 V and a rated current of 14 A. Therefore, it needs a charge controller rated at least 14 A. However, owing to various environmental factors, the ampacity level may increase. For this reason, an additional 25% current should be factored in. Thus, the minimum ampacity required for this charger rises to $14 \times 1.25 = 17.5$ A. So, the solar module would now require a 12 V and a rounded up 20 A charge controller.

Using charge controllers with a higher ampacity is a good practice as it will not harm the system and will be helpful if the size of the PV system is increased in the future.

5.4.11.1 PWM Charge Controller Sizing

PWM controllers cannot limit their output current and only use the array current. So, if the charge controller is rated at 40 A and the solar module is producing 50 A of current, the controller may get damaged. It is important to make sure that the charge controller is compatible with the module and sized properly. A PWM controller with a rating of 40 A can handle up to 40 A current. It is important to check the amperage and voltage rating of a PWM controller. The nominal system voltage indicates the battery voltage that the controller can handle. So, the rated voltage of the controller and the battery bank should match. To consider the rated battery current, let us look at an example.

Example 5.4 (PWM Charge Controller Sizing) Assume four 100 W modules in parallel where each one of them has I_{SC} of 5.86 A. These modules are feeding a 200 W (12 V) load. The sizing of the PWM controller is calculated as follows.

The total short-circuit current is $4 \times 5.86 = 23.44$ A. The charge controller should withstand 125% of the total I_{SC}. The minimum current rating of the PWM controller is $23.44 \times 1.25 = 29.3$ A.

The charge controller should have an ampere rating of more than 29.3 A. For a 12 V system, there are several PWM charge controllers available. Therefore, for this system, the SRNE PWM HP 24 V 30 A charge controller can be selected. A typical list of parameters for the available PWM charge controllers in the market is given in Table 5.5.

Now, let us make sure that the output current rating is greater than the maximum load current. The maximum load current can be calculated as

$$\frac{200\ W}{12\ V} = 16.67\ A.$$

Therefore, the controller of choice satisfies withstand both maximum I_{SC} with 125% safety factor and maximum load current.

The table shows that the controller is suitable for the 12 V, 30 A systems. The maximum solar input represents the amount of voltage that is allowed to go into the controller. In this case, the controller can accept a maximum of 48 V for an 18 V system. Each controller used in the system has a minimum gauge size for the

Table 5.5 Parameters for PWM charge controller sizing

Model	Parameter
Nominal system voltage	12 V/24 V auto-recognition
Rated charge current	30 A
Maximum solar input voltage	48 VDC
Terminals	Up to #8 AWG
Battery type	Sealed (absorbent glass mat (AGM)) gel, flooded, and Li-ion

Table 5.6 Parameters for MPPT charge controller sizing

Model	ROV 20	ROV 40
Rated battery current	20 A	40 A
Nominal system voltage	12 V/24 V auto-recognition	
Maximum solar input voltage	100 V_{DC}	
Battery type	Sealed (AGM), gel, flooded, and Li-ion	

terminal, which is also important to consider during wire sizing for the system. In this example, an #8 AWG wire will suffice. Finally, the type of batteries must be considered, which should be compatible with the charge control unit or not.

5.4.11.2 MPPT Charge Controller Sizing

MPPT controllers have the ability to limit their output, and so they are used where the solar array voltage is higher than the battery bank voltage. This results in less efficiency of the system. However, even if the modules generate 70 A of current, an MPPT controller rated at 40 A will always provide 40 A of output current. The voltage rating of an MPPT controller is higher than the battery, unlike PWM controllers. This is because of the MPPT controller's special ability to lower the voltage and enhance the current to compensate for the power loss. An example of MPPT controller sizing is given below.

Example 5.5 (MPPT Charge Controller Sizing) In this example, the MPPT charge controller sizing method is explored.

ROV is an acronym for the rover series MPPT charge controller. For example, if the ROV value is 40, it means the controller is rated for 40 A current. The MPPT controller can handle 12 V battery voltage. So, now the ROV value should be checked. Table 5.6 shows the required parameters for the sizing of the MPPT charge controller.

Next, the maximum solar input voltage is taken into account, which in this case is 100 V. Therefore, the MPPT controller can receive the input up to 100 V and step it down for the 12 V or 24 V battery. For example, suppose four 100 W modules are connected in series (a string of 400 W), and their open-circuit voltage is 22.5 V (each). So, the total voltage for those 4 modules would be

$$22.5\,V \times 4 = 90\,V.$$

This is the amount of voltage the controller can accept. In order to calculate the output current of the charge controller, simply divide the wattage by the voltage. In this example, the output current for a 12 V battery would be

$$\frac{400\,W\ solar\ array}{12\,V\ battery\ bank} = 33.33\,A.$$

This means that the rating of the charge controller should be 40 A at least as the final output will be multiplied by a safety factor of 1.25, which will result in

$$33.33 \times 1.25 = 41.6625\,A.$$

It is okay to slightly oversize a charge controller where it will face higher energy than it is rated for, as long as it is within the accepted range set by the manufacturer. During the highest solar irradiation hours, the module produces the peak output. The charge controller clips the output to limit it to 40 A. But, when the output is less than 40 A during the rest of the day, the charge controller sends out the complete output. To learn more about sizing charge controllers, visit Refs. [37, 38].

5.5   Inverter

An inverter is a power electronic device that converts DC power into AC power at a specific voltage and frequency. Most electrical devices, such as fridges, dishwashers, lighting, and heating devices, run on AC power. On the other hand, a solar PV system outputs DC power. So, it is necessary to introduce power conditioning units to the PV system for DC–AC power conversion. Thanks to the advanced power electronics development in the last few years, inverters' roles have succeeded in doing the DC–AC conversion process and control the charging and discharging processes of batteries. Moreover, inverters help to obtain maximum output from the PV modules using MPPT techniques. In addition, they control the speed and rotational direction of motors by changing the frequency and voltage of the output AC power to reduce the inrush starting current of the induction motors and enhance motors speed control. This control technique is called variable frequency drive (VFD), a technology widely deployed in solar water pumping applications. Inverters can be of two main types: off-grid and grid-tied inverters.

5.5.1   Battery-Based or off-grid Inverter

A battery-based inverter, also known as an off-grid inverter, is a DC–AC power converter that does not integrate with the grid. Instead, it supplies power to the load directly by converting DC power into AC, taking from the battery. Battery-based inverters can function in off-grid PV systems, but some of the types can also send power back into the utility grid. Such inverters are known as hybrid inverters, which require a battery system to operate. Off-grid inverters are classified into three different types based on the output voltage shape. The output waveforms of the different types of inverters are illustrated in Fig. 5.32:

- **Square wave inverter:** It is the simplest and most inexpensive inverter that has minimal applications. The output signal can be formed by a switch that permits

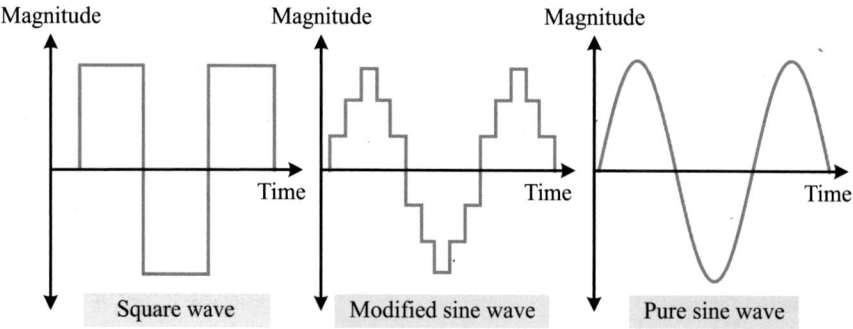

Fig. 5.32 The output waveform of different types of inverters

current flow in one direction during the first half cycle and then reverses the direction in the second half cycle.

- **Modified sine wave inverter:** It is also called modified square wave inverter. It is based on a complex technology compared to the square wave inverter. The inverter can generate multilevel voltage (or current) values using PWM technology. The modified square wave generates a similar wave to the pure sine wave. However, it is not recommended to use this inverter to run an induction motor due to its high value of harmonics, which can easily burn the motors after a few operations.

- **Pure sine wave inverter:** It has the same technology as the modified sine wave inverter but also uses some inductors and capacitors as filters. Filters are required to remove the unwanted components from the signal and make it as ideal as possible. These are used to filter out the harmonics and smooth the output signals to match the original utility signal. These types of inverters are widely used in developing countries due to the long and unexpected power shortages. In addition, pure sine wave inverters can be utilized as a master inverter for several input resources (hybrid inverter) that can have several inputs (such as from solar arrays, batteries, utility grid, and a diesel generator) and two outputs (such as AC loads and batteries) as shown in Fig. 5.33.

Not all battery-based inverters are, however, off-grid. Due to the increasing addition of renewables into the grid, and the necessity of energy storage for a reliable energy source and other applications, batteries can also be used for grid-side applications, which necessitate the use of battery-based inverters for utility-scale applications too. So, some battery-based inverters can be grid-tied too.

5.5.2 Grid-Tied Inverter

In a solar PV system, a grid-tied inverter is used between the solar arrays and the grid. It performs DC–AC power conversion and feeds the utility grid or the

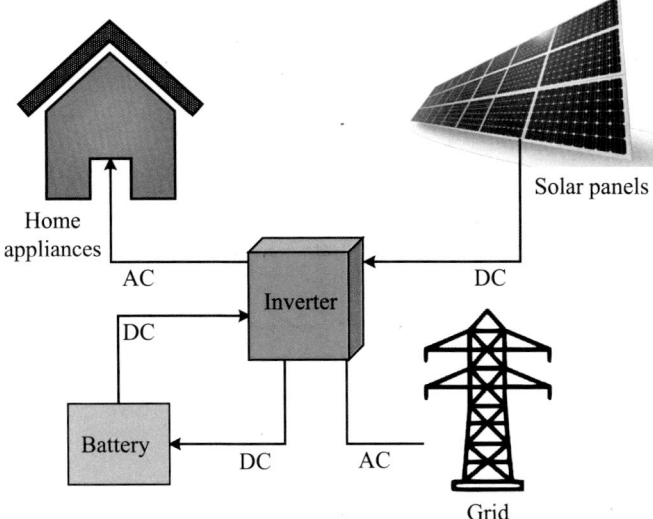

Fig. 5.33 Connection of input–output of hybrid inverters

connected AC loads. To safely carry out this operation, the inverter output should have the same frequency, phase sequence, line voltage, phase angle, and waveform as the grid to meet the criteria of grid synchronization.

Since the main idea of grid-tied inverters is to integrate solar PV energy into the conventional grid, it is obvious that neither the square wave inverter nor the modified sine wave inverter is able to maintain the quality of the power, except the pure sine waver inverter. The classification of grid-tied systems is based on size and system configuration, as shown in Fig. 5.34. Some solar PV systems feed their entire generation to the grid, and the grid feeds the loads separately. The other solar PV systems support their connected loads first, while the BESS or the grid compensates for power shortages.

5.5.3   Inverter Sizing

In a solar PV system, DC power is provided by the solar modules, which is converted into AC power by the inverter, and then it is fed to the load/grid. So, the lesser the inverter power loss, the more efficient it will be. For this reason, in order to use an inverter, the required size should be determined for a specific load. The inverter should have a nominal voltage equal to the battery, and the input rating should be higher than the total power rating of appliances. The solar inverter comes in different sizes and can be rated in VA and Watts. For a standalone system, the inverter must be large enough (typically 20–30% bigger) to handle the total power of the appliances and to maintain a safety margin for a possible load expansion. For grid-connected

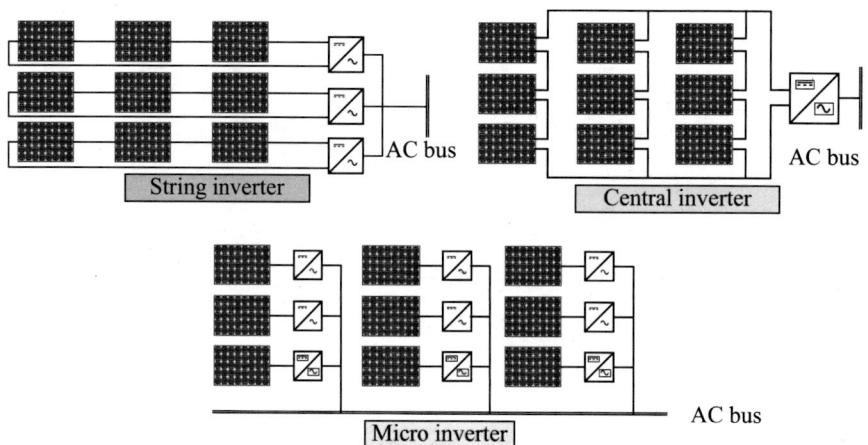

Fig. 5.34 Grid-tied inverter configurations

systems, the PV array capacity should be within the accepted range specified by the inverter's manufacturer in the datasheet. Grid-tied inverters can be connected to a higher solar module capacity than their AC power rating. For instance, SMA Sunny Central Inverter has an AC power rating of 3 MW, while it can be connected to 4.8 MW of solar modules [39].

Example 5.6 (Inverter Sizing) In this example, the inverter sizing method is illustrated in the following NEC standards.

Suppose a load requires a total power of 535 W. To run the load, a typical 12 V battery needs to provide a current of

$$\frac{535\ W}{12\ V} = 44.6\ A.$$

Now, an inverter's required capacity (VA rating) needs to be measured. In an ideal condition, an inverter would operate with 100% efficiency, meaning all the powers provided by the battery would be inverted for further usage. However, in reality, inverters have an efficiency of 95–98%, i.e., they consume some portion of the power, while the conversion process occurs.

Let us assume that the efficiency of the inverter is 98%. So, the current that will be provided is

$$\frac{44.6 A}{0.98} = 45.51\ A. \tag{5.17}$$

For this example, a power factor of less than 100% is observed, and so, power factor compensation of 0.8 is considered. Therefore, to run the inverter and support the load, the battery is required to increase the current flow. The required VA of an inverter is the apparent power sent from the inverter to the load, which can be found from the following equation:

$$Inverter\ VA\ rating = \frac{Total\ wattage}{Power\ factor} = \frac{535}{0.8} = 668.75 \approx 669\ VA. \quad (5.18)$$

So, an inverter with 700 VA that is available in the market would be the right choice for usage. To compute the battery backup time for this particular inverter, the calculations are done as per Eq. 5.19.

$$Battery\ backup\ time$$
$$= \frac{Battery\ voltage,\ V \times Battery\ capacity,\ Ah \times No.\ of\ batteries}{Connected\ load,\ W}.$$
$$(5.19)$$

In this example, let us assume two 12 V batteries, each having a capacity of 200 Ah, and they are used in series. Using Eq. 5.19, the battery backup time would be

$$\frac{12\ V \times 200\ Ah \times 2}{535\ W/0.98} = 8.79\ h.$$

Here, 0.98 has been divided to show that the inverter efficiency is 98%. So, the battery can supply power for 8.79 h. To learn more about inverter sizing, visit Refs. [40, 41]. In consideration of other factors such as the ambient temperature, DoD of the battery, inverter efficiency, and the age of the components, the calculation can be more complex than the one in Example 5.7.

5.6   Balance of System Components

In a solar PV system, all the components except the PV arrays may be considered as the balance of system (BOS) components. Such components include the inverter, battery, and charge controller as well, but considering the importance and large size of these components, they have been separately treated in the preceding sections. The BOS components covered in this section are the wiring, overcurrent and lightning protection devices, switches and disconnects, grounding, mounting, tracking systems, and cooling systems. Some other components often used in a PV system are net energy meters, battery management system (BMS), solar concentrator, pyranometer, albedometer, solar irradiance sensors, and so on. These are not covered in this book due to the concise nature of the book.

5.6.1 Wiring

The wiring process is done by conductive wires surrounded by an insulator, such as polyvinyl chloride (PVC), to protect the users and the surroundings from any electrical shock. The conductive core of the cable consists of a solid metal wire or multiple tiny soldered wires, which are preferable over the solid one due to the flexibility. These metallic conductors are usually made from Aluminum or Copper. Cost, conductivity, and availability are the three factors determining the type of wires. Copper has a higher price due to its lower resistivity than Aluminum; the resistivity of Copper and Aluminum are $1.68 \times 10^{-8}\ \Omega m$ and $2.65 \times 10^{-8}\ \Omega m$, respectively. In other words, Copper is 160% more conductive than Aluminum.

In order to select the suitable wire, the wire of choice must handle the highest possible current that might pass through the circuit. In addition, it must have low resistance in order to avoid a fixed percentage of allowed voltage drop (this will be further explained in Sect. 5.6.1.4).

5.6.1.1 Ampacity
The maximum amount of current (A) carried by the conducting wires under the rated operating temperature is known as ampacity. The ampacity is a portmanteau for ampere capacity, which is measured in Ampere. The ampacity is strongly related to the electrical cable resistance, R, which is shown in Eq. 5.20.

$$R = \rho \frac{l}{A},\tag{5.20}$$

where

ρ = resistivity of the cable material (Ωm),
l = length of the material (m), and
A = Cross-section area of the cable (m^2).

The above units are provided in the SI units, but they can also be represented in terms of Ω feet, feet, and feet^2, respectively.

Since the resistivity of the wire material is a constant value, and the length varies based on the application, wires are distinguished based on their sizes or cross-sections. The higher the cross-section area, the more current the cable can carry. In Europe, the wires are measured in mm^2, where the values start from $1\ mm^2$ up to $1000\ mm^2$. However, in the USA, the wire sizes are measured in American Wire Gauge (AWG). To convert from AWG into mm^2, Table 5.7 can be used. Notice that the AWG numbers increase as the cross-sectional area in mm^2 decreases, i.e., the wires get thinner. Another unit commonly used in the North American electrical industry is thousands of circular mils, or kcmil in short. The conductors larger than size 4/0 AWG are generally identified by the area in kcmil, where 1 kcmil = $0.5067\ mm^2$. A circular mil is the area of a wire whose diameter

Table 5.7 Conversion table of AWG into mm^2 units of wire sizing [42]

AWG	mm^2	AWG	mm^2	AWG	mm^2	AWG	mm^2
4/0	107	8	8.37	19	0.653	30	0.0509
3/0	85	9	6.63	20	0.518	31	0.0404
2/0	67.4	10	5.26	21	0.41	32	0.032
1/0	53.5	11	4.17	22	0.326	33	0.0254
1	42.4	12	3.31	23	0.258	34	0.0201
2	33.6	13	2.62	24	0.205	35	0.016
3	26.7	14	2.08	25	0.162	36	0.0127
4	21.2	15	1.65	26	0.129	37	0.01
5	16.8	16	1.31	27	0.102	38	0.00797
6	13.3	17	1.04	28	0.081	39	0.00632
7	10.5	18	0.823	29	0.0642	40	0.00501

is one mil. One million circular mils (MCM) refers to the area of a circular cross-section whose diameter is 1000 mil or 1 in.

5.6.1.2 Wire Types

Comparison Between DC Cable and AC Cable
A PV system requires both DC and AC cables. The wires from the PV module to the charge controller, from the charge controller to the battery bank, and from the battery to the inverter input are DC cables. On the other hand, the wires from the inverter output to the load/grid are AC cables. Therefore, when purchasing the wires, this difference in the cable types should be kept in mind. In this section, the differences between AC and DC cables are explained.

The key difference between DC and AC cables lies in the amount of insulation surrounding the conductor. The cable insulation is directly dependent on the working voltage of the power flowing through the conductor. So, the higher the voltage rating, the higher and thicker will be the insulation required.

DC voltage refers to the peak value of the voltage, whereas AC voltage refers to the RMS value of the voltage. The relationship between peak and RMS voltages can be expressed using the relationships in Eqs. 5.21 and 5.22.

$$V_{peak} = \sqrt{2} \times V_{rms}, \tag{5.21}$$

$$V_{rms} = 0.707 \times V_{peak}. \tag{5.22}$$

So, when we say that the voltage of an AC system is 120 V, the system actually experiences a higher voltage during the peak of the AC cycle. Therefore, the system is designed to withstand about $120 \times \sqrt{2} \approx 170$ V of peak voltage. Since the RMS value is 70.7% of the peak value, so AC cables require 29.3% more insulation compared to DC cables for the same working voltage.

In AC cables, a phenomenon known as the skin effect occurs, which is absent in DC cables. When AC power flows through a wire, it has a tendency to flow close to the surface of the wire. This is called the skin effect. Since the current density is the highest close to the surface of the wire, so the effective cross-section of the wire diminishes. Thus, the resistance of the wire, being inversely proportional to the cross-sectional area (Eq. 5.20), gets increased. So, the skin effect causes the resistance of AC cables to be higher than DC cables. Thus, the power losses in AC cables are higher than in DC cables. The higher the AC frequency, the higher will be the skin effect, and the higher will be the resistance, thereby increasing the losses. In addition, losses in AC wires are incurred due to the effect of power factor, inductance, capacitance, etc. So, the overall losses are much higher in case of AC compared to DC.

Therefore, due to higher power losses and higher insulation required, the cost of AC wires is generally higher compared to DC wires. However, the overall cost comparison cannot be based on any single factor. For any specific application, a thorough evaluation of the available choices and the system variables will determine which type of power make is more expensive.

Cable Materials

The wires used in electrical systems are usually made of copper, aluminum, copper-clad aluminum, nickel, or nickel-coated copper. Copper and aluminum are the two most widely used conductor materials. The conductivity of copper is about 1.63 times higher than that of aluminum, but aluminum is lighter in mass than copper [43]. With the same value of resistance, a bare copper wire weighs double compared to a bare aluminum wire. Besides, copper is more expensive than aluminum. The tensile strength of copper is about 40% higher than aluminum [44], which implies that copper wires are less likely to break due to strain, making it more convenient to pull the wire through ports and feeders. When exposed to heat, the thermal expansion of copper is lower compared to aluminum. This is another indicator of the fragility of aluminum as a conductor since an expanded wire can tear more easily.

Therefore, there are downsides to both copper and aluminum wires, as well as their own advantages. For residential uses, copper wires are more popular than aluminum for their strength and conductivity. In the examples throughout this book, copper wires are predominantly used.

Cable Naming

The cable names contain some specific letters, each of which bears a special meaning. The significance of the letters used in cable naming is defined in Table 5.8. Table 310.4(A) from NEC 2020 [45] specifies the application provisions, operating temperature, insulation type and thickness, and other important characteristics of different types of conductors. A small excerpt from the table is presented in Table 5.9.

A different approach to naming cables is to name the cables based on the number of conductors or cores encased within the insulating layer. In this convention, 1/C

Table 5.8 Definition of the letters typically used in cable naming

A	Asbestos
E	Entrance
F	Feeder
FEP	Fluorinated ethylene propylene insulation
H	Heat resistant
HH	High heat resistant
N	Nylon-coated
PFA	Perfluoroalkoxy insulation
R	Rubber
S	Service or Silicone
T	Thermoplastic
U	Underground
W	Water-resistant
X	Cross-linked polyethylene (XLPE) insulation
Z	Ethylene tetrafluoroethylene (ETFE) insulation
-2	90 °C temperature rating

Table 5.9 The characteristics of the most popular cables [45]

Acronym	Full name	Location	Temperature, °C (°F)
RHH	Rubber high heat resistant	Dry and damp locations	90 (194)
RHW	Rubber heat resistant	Dry and wet locations	75 (167)
RHW-2	Rubber heat resistant	Dry and wet locations	90 (194)
TW	Thermoplastic water-resistant	Dry and wet locations	60 (140)
THW	Thermoplastic heat- and water-resistant	Dry and wet locations	75 (167)
THW-2	Thermoplastic heat- and water-resistant	Dry and wet locations	90 (194)
THHW	Thermoplastic high heat- and water-resistant	Wet location	75 (167)
		Dry location	90 (194)
THWN	Thermoplastic heat- and water-resistant nylon-coated	Dry and wet locations	75 (167)
THWN-2	Thermoplastic heat- and water-resistant nylon-coated	Dry and wet locations	90 (194)
THHN	Thermoplastic high heat-resistant, nylon-coated	Dry and damp locations	90 (194)
XHHW	XLPE (cross-linked polyethylene) high heat- and water-resistant	Wet location	75 (167)
		Dry and damp location	90 (194)

cable implies a cable with only one conductor inside, 2/C cable has two separate conductors inside, and 3/C cable has three separate conductors inside. 3.5/C and

Table 5.10 A general guide
to equipment voltage ratings
according to industry
standards

Name	Voltage rating
Standard voltage	0–600
Low voltage	601 V to 2 kV
Medium voltage (MV)	2.001–35 kV
High voltage (HV)	Above 35 kV

Table 5.11 A general guide
to distribution system voltage
ratings according to industry
standards

Name	Voltage rating
Low voltage	0–49 V
Medium voltage (MV)	50 V to 1 kV
High voltage (HV)	Above 1 kV

4/C cables are also available. It is noteworthy that multi-core cables are different from multi-stranded cables, which have multiple strands within one core.

Cable Voltage Rating
Electrical cables may be rated in terms of their RMS voltage in the case of AC systems or the highest system voltage in the case of DC systems. The cables can be classified as shown in Table 5.10, although the values may differ from place to place. PV systems are usually LV systems, and therefore, all the systems described in this book are LV systems by default.

The voltage rating classification is differently done for distribution systems, which is shown in Table 5.11.

5.6.1.3 Wire Sizing Tables—NEC 2020 [45]
The National Fire Protection Association (NFPA) has several codes, standards, and handbooks for safe electrical systems. The NFPA 70 Handbook is named the National Electric Code (NEC), and alongside many other aspects of electrical system design, it provides several instructions and tables for wire sizing of several systems. You can freely access the NEC 2020 Handbook from NFPA's official website.[1] You need to register an account to be able to access the handbook. It is noteworthy that in the USA, the foot-pound-second (FPS) system of units is preferably used, which is why the NEC uses the FPS system. Change the units as required if you are using a different standard of measurement.

The NEC instructs the allowable ampacity of the available wire sizes of different conductor's materials (Copper and Aluminum or Copper-clad Aluminum) and different temperature ratings (60 °C, 75 °C, 90 °C) in two possible criteria:

1. Insulated conductors with not more than three current-carrying conductors rated up to 2 kV in raceway, cable, or earth, given in Table 310.16 of NEC 2020. For

[1] https://www.nfpa.org/Codes-and-Standards/All-Codes-and-Standards/Free-access.

example, if we choose THHW Copper conductor at 75 °C, we need to use #6 AWG for 65 A. Again, for 115 A ampacity, we will require #2 AWG.
2. Single-insulated conductors rated up to 2 kV in free air in Table 310.17 of NEC 2020. For instance, if we choose THHW Copper conductor at 75 °C, we need to use #6 AWG for 95 A. Again, for 170 A ampacity, we will require #2 AWG.

These ampacity values are correct under the ambient temperature of 30 °C (86 °F) (or 40 °C in other cases). Otherwise, for different ambient temperatures or for cables with more than 3 insulated conductors, the correction factors provided in the two tables must be used in order to obtain the new ampacities correlated with the expected operating conditions of the designed system. The two tables are described below:

- Table 310.15(B)(1) refers to the ampacity correction factor based on the ambient temperature of 30 °C, for temperature ratings 60 °C, 75 °C, or 90 °C. For example, if the ambient temperature is 36–40 °C, and if the temperature rating of the conductor is 75 °C, then the rated current should be multiplied by a factor of 0.88. A similar table is Table 310.15(B)(2), which is based on the ambient temperature of 40 °C.
- When more than 3 conductors are used, the ampacity decreases from the rated value by a certain percentage. Table 310.15(C)(1) enlists these adjustment factors where more than three current-carrying conductors are used. For instance, if the number of conductors is between 4 and 6, then the current rating has to be multiplied by a factor of 80%, and if the number of conductors is between 7 and 9, then the current rating has to be multiplied by a factor of 70%.

Suppose one or two of the conditions do not match with the standards. In that case, the designer will have to multiply the correction factor of the operating condition by the original ampacity given in Table 310.16 (NEC 2020) to calculate the new ampacity. For any increase in the ambient temperature or the number of adjacent conductors, derating will occur in the conductor's ampacity.

Ampacity reflects the current-carrying capacity of a conductor, and it varies depending on natural variables such as temperature. The upcoming two examples illustrate the calculation of conductor ampacity.

Example 5.7 (Wire Sizing for Different Ambient Temperatures) This example deals with the relation between temperature and conductor ampacity.

Say we need to determine the ampacity of a #4 TW Copper conductor whose temperature rating is 60 °C but placed at an ambient temperature of 42 °C. According to Table 310.16, its ampacity is 70 A. But since the ambient temperature is not 30 °C, the ampacity should be multiplied by 0.71, according to Table 310.15(B)(1). So, the new ampacity of the conductor is $70 \times 0.71 = 49.7$ A. Thus, if the ambient temperature of the cable increases, then the ampacity of the conductor reduces.

Example 5.8 (Wire Sizing for Multiple Wires Inside Raceway) Suppose we need to find the ampacity of a #2 THW Copper wire whose temperature rating is 75 °C. According to Table 310.16, its ampacity is 115 A. But if the cable has 4 wires inside the raceway, then we need to multiply 115 A by 0.8 according to Table 310.15(C)(1). So, the new ampacity of the wire is $115 \times 0.8 = 92$ A. Thus, if the number of wires inside the cable increases, then the ampacity of each conductor reduces.

Let us take a look at another example. Suppose a string has a short-circuit current rating of 12.20 A. Assuming all the terminals are rated at 75 °C. To calculate the conductor ampacity for the PV source circuit, NEC sections 690.8(A)(1) and 690.8(B)(1) are considered, which suggest multiplying the short-circuit current and current rating of the conductor by 125%, respectively. So, the ampacity of the conductor is determined using the following equation:

$$
\begin{aligned}
Ampacity\ of\ the\ conductor &= (Module\ I_{SC} \times 1.25) \times 1.25 \\
&= (12.20\ A \times 1.25) \times 1.25 \qquad (5.23) \\
&= 19.06\ A.
\end{aligned}
$$

Article 690 in the NEC 2020 Handbook discusses solar PV systems. Two statements are noteworthy in this regard:

- As per Article 690.8(A)(1) of the NEC 2020, the maximum current of a PV source circuit will be 125% of the sum of the rated I_{sc} of the PV modules connected in parallel. This criterion ensures that the wire does not exceed $1 \div 1.25 = 80\%$ of its nominal ampacity so that the wire has adequate protection against overheating due to the flow of current.
- As per Article 690.8(B)(1) of the NEC 2020, the current rating of the circuit conductors should be multiplied by 125% of the rated capacity before any adjustment or correction factors are applied.

According to the handbook, when both Articles 690.8(A)(1) and 690.8(B)(1) of NEC 2020 are applicable, a 156% safety factor ($1.25 \times 1.25 = 1.56$) should be multiplied by the current or the sum of the currents that are directly related to the solar irradiance. These currents can be limited to the currents in the PV source circuit and PV output circuit.

PV Source Circuit This circuit is made of series-connected PV modules, which is also called a string. This string is connected to a common point known as the DC combiner box, which combines all the outputs from the PV modules.

PV Output Circuit The connection point of the PV source circuits usually takes place at a DC combiner box, where the output of this box is called a PV output circuit. The output is mainly the sum of the parallel-connected PV source circuit currents.

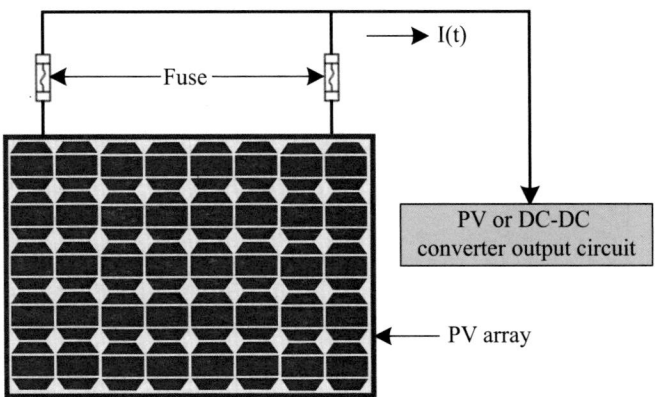

Fig. 5.35 PV array components [45]

The safety factor 156% consists of two 125% factors ($1.25 \times 1.25 = 1.56$). The first factor guarantees that the wire does not exceed 80% of its nominal ampacity. The second factor is related to any unfavorable atmospheric condition that might let the PV modules produce more than its rated currents. This situation might be related to the altitude of the installation site, where the irradiance might exceed $1000\,W/m^2$, which is listed as a standard test condition. It might also occur during a sunny day on a surface covered by snow that has perfect reflectivity to the modules. It means more current than the one listed in the datasheet. Based on this, the component arrangements of the PV array are shown in Fig. 5.35.

Besides, there are many other tables that are needed for wire sizing. For example, Table 8 (in Chapter 9 of NEC 2020) enlists the conductor properties such as size, diameter, area, and resistance for coated and uncoated Copper or Aluminum conductors. For instance, the resistance of an uncoated copper wire of #6 AWG is $0.491\,\Omega/kft$. Table 9 in the same chapter provides a list of the AC resistance and reactance for 600 V cables, 3 phase, 60 Hz, 75 °C (167 °F) with three single conductors in a conduit.

5.6.1.4 Voltage Drop

The voltage drop is an electrical loss that occurs as electrical potential reduction along the electrical conductor through which current is passing. Voltage drop can be simulated as a single resistor representing the cumulative resistance of the whole length of the wire. Voltage drop is recommended to be less than 3% between the PV modules and the inverter, as per NEC 2020.

According to Article 215 of NEC 2020, in a 2-wire DC or AC circuit, or a 3-wire AC circuit connected to a balanced load, with negligible reactance and 100% power factor, the voltage drop can be given by Eq. 5.24.

$$V_D = \frac{2 \times L \times R_C \times I}{1000},$$

$$(5.24)$$

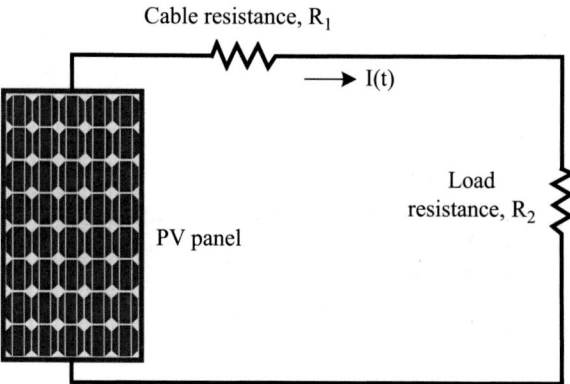

Fig. 5.36 Voltage drop equivalent circuit

where

V_D is the voltage drop based on the conductor temperature of 75 °C,
L is the one-way length (feet) of the circuit,
R_C is the conductor resistance (Ω per 1000 ft) taken from Table 8 of Chapter 9 in
    NEC,
and I is the current (A) flowing through the circuit.

The voltage drop equivalent circuit is illustrated in Fig. 5.36. The voltage drop
across R_1 can be calculated using Eq. 5.25.

$$V_{R_1} = V_{source} \times \frac{R_1}{R_1 + R_2}. \tag{5.25}$$

In Eq. 5.25, the lower the R_1 value, the lower the voltage across the cable.
Therefore, in order to reduce the voltage drop, the system designer needs to increase
the wire sizing until the ratio V_{R_1} : V_{source} is less than the specific percentages
mentioned above. The voltage drop is initially related to three parameters:

- **Wire size or cross-section:** The greater the cross-section, the lower the voltage
  drop as the resistance is lower.
- **Length of wire:** The longer the wire, the higher the voltage drop, since the wire
  resistance is proportional to the length of the wire.
- **Current:** The current is also proportional to the voltage drop, since $V_{R_1} = R_1 \times I$.

Neglecting the effect of voltage drop in the wire sizing process might lead to a
failure in the system, especially if the terminal voltage of solar PV systems falls
below the inverter threshold level. Voltage drop also has a dreadful consequence

Table 5.12 Electrical characteristics of PV module

Characteristics (STC)	Value
Open-circuit voltage (V_{oc}), V	37.51
Short-circuit current (I_{sc}), A	8.7
Maximum power voltage (V_{mpp}), V	30.42
Maximum power current (I_{mpp}), A	8.22
Maximum power (P_{max}), W	250
Module efficiency (η), %	15.24

in submersible water pumping applications. The pump's motor can be as deep as 400 m, and voltage might be significantly dropped along the long wire, preventing the motor from receiving the nominal operating voltage.

5.6.1.5 Wiring Sizing Example
Wire sizing calculation is very important to ensure safe and reliable operation and working at full load without being damaged. Proper sizing allows a cable to withstand overcurrent and short-circuit current. An example of wire sizing with a detailed explanation and calculation is given below.

Example 5.9 (Wire Sizing Example for PV System) Twelve modules having a nominal power rating of 250 W each form a solar array to charge eight (200 A/12 V) lead-acid batteries. The batteries are connected to a pure sine wave, 3 kW inverter with a nominal output current of 12.5 A. The solar array contains 4 power circuits (strings), each with three modules in series, whereas the battery system voltage is 48 V. The PV module specification is provided in Table 5.12.

Some values that are required for sizing are:

1. For a less than 2% voltage drop on the DC side and less than 1.5% voltage drop on the AC side, the maximum length of an uncoated #6 AWG copper wire (48 V, 95 A) is 10.29 ft as given in Table 5.15 [46].
2. For 75 °C THHW type of copper wire, #1 AWG can handle 130 A, #6 AWG can handle 65 A, and #14 AWG can handle 20 A as given in Table 310.16 of NEC 2020.
3. The DC resistance at 75 °C of copper wire of size #1 AWG is 0.154 Ω/kft, and #6 AWG is 0.491 Ω/kft as given in Table 8, Chapter 9 of NEC 2020. The DC resistance from Table 8 will be necessary for calculating the voltage drop in the DC circuit.
4. The AC resistance at 75 °C of copper wire of size #14 AWG is 2.7 Ω/kft as given in Table 9, Chapter 9 of NEC 2020. The AC resistance from Table 9 will be necessary for calculating the voltage drop in the AC circuit.

First, calculate the wire size between the PV array and the batteries and then between the battery bank and the inverter, considering the ampacity and voltage drop requirements. Then, utilize the following information provided by the solar PV module manufacturers:

- The temperature will not exceed 30 °C.
- The distance between the PV modules and the batteries is 100 ft.
- The distance between the batteries and the inverter is 7 ft.
- The distance between the inverters and the AC distribution panel is 30 ft.
- The efficiency of the inverter is 95%.

Wire Between the PV Modules and the Batteries

Step 1:   **Determination of wire size as per current rating.** First, let us start with calculating the sum of the currents provided by the parallel solar circuits using the following equation:
Total current = each power source I_{sc} × number of power sources of the array.
or, total current = 8.7 A ×4 = 34.8 A.
The wire should not be more than 80% loaded, as suggested in Article 690.8(B)(1) from NEC 2020. Therefore, the total current should be multiplied by 125% or divided by 80%. Then, another 125% must be multiplied to handle any possible value of produced current that might exceed the standard test conditions (STC) nominal current.
∴ Total current with safety factors = 1.25 × 1.25 × 34.8 A = 54.375 A.
Assuming that THHW (75 °C) wire type is available in the market, by getting Table 310.16 (NEC 2020), a segment of which is given in the question, the chosen size is #6 AWG for copper wire. This wire size handles a current of 65 A, which is the closest higher number to 54.375 A in the table. We need to check if this meets the voltage drop requirements, which must be below 2%.

Step 2:   **Determination of voltage drop within the chosen wire.** In this example, the V_{mpp} of one module = 30.42 V.
So, V_{mpp} of the string = V_{mpp} of one module × number of PV modules in series = 30.42 V ×3 = 91.26 V.
For the DC side, the voltage drop in the wire should not exceed 2% of this value, i.e., 1.8252 V. For the AC side, the voltage drop in the wire should not exceed 1.5% of this value, i.e., 1.3689 V. To ensure that the voltage drop within the wires is within a tolerable range, the determination of the total voltage of the PV module is necessary.

The equation for determining the voltage drop within any wire can be derived from Eq. 5.24 and is as follows:

$$V_{drop} = I \times R_c \times L, \tag{5.26}$$

where

I = the maximum operating or load current, in Ampere;
R_c = the resistivity of the wire, measured in Ω/kft;
L = the total length of the cable, in kilo feet (kft);

Note that the value of I can be found in two ways:

1. Each power source $I_{sc} \times$ number of power sources of the array $\times$ 125%.
2. Each power source $I_{mpp} \times$ number of power sources of the array $\times$ 125%.

Both cases contain just one safety factor instead of two, as seen in the ampacity requirements. The first way is considered to be more conservative. However, both of them are totally fine.

The value of R_c can be found in either Table 8 or Table 9 in Chapter 9 of NEC 2020. If the wire is chosen to carry DC current, Table 8 has to be used, and if it is chosen to carry AC current, Table 9 has to be used.

L is the distance between the PV modules and the batteries. Generally, the two-way wiring requires the distance to be multiplied by 2 to get the total length of the wire.

Therefore, the voltage drop for the #6 AWG wire size will be

$$V_{drop} = (125\% \times 8.22 \times 4) \times 0.491 \times (0.2 \, kft) = 4.036 \, V.$$

$$\frac{V_{drop}}{V_{mpp}} = \frac{4.036}{91.26} = 4.42\%.$$

This value is greater than 2%. Therefore, we need a greater cross-section to decrease the voltage drop.

Step 3:   Choosing another wire size if the voltage drop is too much.
With #1 AWG copper wire, the voltage drop will be

$$V_{drop} = (125\% \times 8.22 \times 4) \times 0.154 \times (0.2 \, kft) = 1.265 \, V.$$

$$V_{drop}\% = \frac{V_{drop}}{V_{mpp}} = \frac{1.265}{91.26} = 1.38\%.$$

So the wire size of choice is #1 AWG copper wire, which can safely handle up to 130 A of current.

Wire Between the Batteries and the Inverter
The maximum continuous power that the inverter can drain from the battery is

$$\frac{Maximum \, inverter \, output \, power}{Inverter \, efficiency} = \frac{3000 \, W}{0.95} = 3158 \, W.$$

As discussed in Sect. 5.3, the voltage provided by the battery varies based on the SoC. Therefore, the maximum current that the wires should handle is the current correlated with the lowest possible voltage below, which the inverter will stop draining current from the battery.

The maximum current drained from the battery is

$$\frac{3158\ W}{43\ V} = 74\ A.$$

Considering the safety factor recommended by NEC 2020, the maximum current drained from the battery becomes 74 A × 1.25 = 92.5 A.

From Table 310.16 of NEC 2020, at 75 °C temperature, a THHW copper wire of size #3 AWG can carry up to 100 A of current. According to Table 8, Chapter 9 of NEC 2020, the DC resistance for a copper #3 AWG wire at 75 °C is 0.245 ohm/kft. Here, the total distance of wiring between the battery and the inverter is 7 ft × 2 = 14 ft = 0.014 kft. So, the voltage drop for the #3 AWG wire is then calculated as

$$92.5\ A \times 0.245\ ohm/kft \times 0.014\ kft = 0.317\ V.$$

A 2% voltage drop of the 48 V battery bank translates to 0.02 × 48 V = 0.96 V. So, the voltage drop in the #3 AWG wire is well below the 2% limit. Thus, wire size #3 AWG satisfies both ampacity and voltage drop requirements.

Alternate Way Observe Tables 5.13, 5.14, and 5.15. These three tables define the critical length of the wiring when the maximum ampacity is known but the wiring distance is not specified. The critical length is the maximum allowable length of the wiring, exceeding which will cause the voltage drop limit to be crossed. In the example here, the distance between the batteries and the inverter is specified as 7 ft. If it were not said, and we were to find out the optimum distance, we could refer to those three tables. This is a 48 V battery bank in question, with a maximum current of 92.5 A. So, the next rounded-up number found from Table 5.15, for a 48 V system, is 95 A. The critical length in the row for the 95 A current has 10.29 ft as the next length after 7 ft (the 7 ft distance is specified at the beginning of this example). So, we are taking 10.29 ft as the critical length here, which corresponds to the wire size #6 AWG. Here, the two-way wiring would involve 10.29 × 2 = 20.58 ft = 0.02058 kft of wiring. From Table 8, Chapter 9 of NEC 2020, the DC resistance for an #6 AWG copper wire is 0.491 ohm/kft. Therefore, for this wire, the voltage drop is

$$95\ A \times 0.491\ ohm/kft \times 0.02058\ kft = 0.96\ V,$$

which agrees with the 2% voltage drop limit of the 48 V system. So, we can conclude that the battery-to-inverter wiring distance in this example may be from 7 ft to 10.29 ft to keep within the voltage drop limits, provided the ampacity is correctly measured.

Table 5.13 Wire lengths, in feet, for different AWG sizes of conductors with a specific ampacity and resistance for a 12 V system [46]. The first column contains the wire ampacity, in Ampere

AWG size	18	16	14	12	10	8	6	4	3	2	1	"1/0"	"2/0"	"3/0"	"4/0"
Copper uncoated DC resistance at 75 °C	7.95	4.99	3.14	1.98	1.24	0.78	0.49	0.31	0.25	0.19	0.15	0.12	0.10	0.08	0.06
1	15.09	24.05	38.22	60.61	96.77	154.24	244.40	389.61	489.80	618.56	779.22	983.61	1240.95	1566.58	1973.68
2	7.55	12.02	19.11	30.30	48.39	77.12	122.20	194.81	244.90	309.28	389.61	491.80	620.48	783.29	986.84
3	5.03	8.02	12.74	20.20	32.26	51.41	81.47	129.87	163.27	206.19	259.74	327.87	413.65	522.19	657.89
4	3.77	6.01	9.55	15.15	24.19	38.56	61.10	97.40	122.45	154.64	194.81	245.90	310.24	391.64	493.42
5	3.02	4.81	7.64	12.12	19.35	30.85	48.88	77.92	97.96	123.71	155.84	196.72	248.19	313.32	394.74
6	2.52	4.01	6.37	10.10	16.13	25.71	40.73	64.94	81.63	103.09	129.87	163.93	206.83	261.10	328.95
7	2.16	3.44	5.46	8.66	13.82	22.03	34.91	55.66	69.97	88.37	111.32	140.52	177.28	223.80	281.95
8	1.89	3.01	4.78	7.58	12.10	19.28	30.55	48.70	61.22	77.32	97.40	122.95	155.12	195.82	246.71
9	1.68	2.67	4.25	6.73	10.75	17.14	27.16	43.29	54.42	68.73	86.58	109.29	137.88	174.06	219.30
10	1.51	2.40	3.82	6.06	9.68	15.42	24.44	38.96	48.98	61.86	77.92	98.36	124.10	156.66	197.37
12	1.26	2.00	3.18	5.05	8.06	12.85	20.37	32.47	40.82	51.55	64.94	81.97	103.41	130.55	164.47
15	1.01	1.60	2.55	4.04	6.45	10.28	16.29	25.97	32.65	41.24	51.95	65.57	82.73	104.44	131.58
18	0.84	1.34	2.12	3.37	5.38	8.57	13.58	21.65	27.21	34.36	43.29	54.64	68.94	87.03	109.65
20	0.75	1.20	1.91	3.03	4.84	7.71	12.22	19.48	24.49	30.93	38.96	49.18	62.05	78.33	98.68
25	0.60	0.96	1.53	2.42	3.87	6.17	9.78	15.58	19.59	24.74	31.17	39.34	49.64	62.66	78.95
30	0.50	0.80	1.27	2.02	3.23	5.14	8.15	12.99	16.33	20.62	25.97	32.79	41.37	52.22	65.79
35	0.43	0.69	1.09	1.73	2.76	4.41	6.98	11.13	13.99	17.67	22.26	28.10	35.46	44.76	56.39
40	0.38	0.60	0.96	1.52	2.42	3.86	6.11	9.74	12.24	15.46	19.48	24.59	31.02	39.16	49.34
45	0.34	0.53	0.85	1.35	2.15	3.43	5.43	8.66	10.88	13.75	17.32	21.86	27.58	34.81	43.86
50	0.30	0.48	0.76	1.21	1.94	3.08	4.89	7.79	9.80	12.37	15.58	19.67	24.82	31.33	39.47

(continued)

Table 5.13 (continued)

AWG size	18	16	14	12	10	8	6	4	3	2	1	"1/0"	"2/0"	"3/0"	"4/0"
Copper uncoated DC resistance at 75°C	7.95	4.99	3.14	1.98	1.24	0.78	0.49	0.31	0.25	0.19	0.15	0.12	0.10	0.08	0.06
55	0.27	0.44	0.69	1.10	1.76	2.80	4.44	7.08	8.91	11.25	14.17	17.88	22.56	28.48	35.89
60	0.25	0.40	0.64	1.01	1.61	2.57	4.07	6.49	8.16	10.31	12.99	16.39	20.68	26.11	32.89
65	0.23	0.37	0.59	0.93	1.49	2.37	3.76	5.99	7.54	9.52	11.99	15.13	19.09	24.10	30.36
70	0.22	0.34	0.55	0.87	1.38	2.20	3.49	5.57	7.00	8.84	11.13	14.05	17.73	22.38	28.20
75	0.20	0.32	0.51	0.81	1.29	2.06	3.26	5.19	6.53	8.25	10.39	13.11	16.55	20.89	26.32
80	0.19	0.30	0.48	0.76	1.21	1.93	3.05	4.87	6.12	7.73	9.74	12.30	15.51	19.58	24.67
85	0.18	0.28	0.45	0.71	1.14	1.81	2.88	4.58	5.76	7.28	9.17	11.57	14.60	18.43	23.22
90	0.17	0.27	0.42	0.67	1.08	1.71	2.72	4.33	5.44	6.87	8.66	10.93	13.79	17.41	21.93
95	0.16	0.25	0.40	0.64	1.02	1.62	2.57	4.10	5.16	6.51	8.20	10.35	13.06	16.49	20.78
100	0.15	0.24	0.38	0.61	0.97	1.54	2.44	3.90	4.90	6.19	7.79	9.84	12.41	15.67	19.74
110	0.14	0.22	0.35	0.55	0.88	1.40	2.22	3.54	4.45	5.62	7.08	8.94	11.28	14.24	17.94
120	0.13	0.20	0.32	0.51	0.81	1.29	2.04	3.25	4.08	5.15	6.49	8.20	10.34	13.05	16.45
130	0.12	0.18	0.29	0.47	0.74	1.19	1.88	3.00	3.77	4.76	5.99	7.57	9.55	12.05	15.18
140	0.11	0.17	0.27	0.43	0.69	1.10	1.75	2.78	3.50	4.42	5.57	7.03	8.86	11.19	14.10
150	0.10	0.16	0.25	0.40	0.65	1.03	1.63	2.60	3.27	4.12	5.19	6.56	8.27	10.44	13.16
160	0.09	0.15	0.24	0.38	0.60	0.96	1.53	2.44	3.06	3.87	4.87	6.15	7.76	9.79	12.34
170	0.09	0.14	0.22	0.36	0.57	0.91	1.44	2.29	2.88	3.64	4.58	5.79	7.30	9.22	11.61
180	0.08	0.13	0.21	0.34	0.54	0.86	1.36	2.16	2.72	3.44	4.33	5.46	6.89	8.70	10.96
190	0.08	0.13	0.20	0.32	0.51	0.81	1.29	2.05	2.58	3.26	4.10	5.18	6.53	8.25	10.39
200	0.08	0.12	0.19	0.30	0.48	0.77	1.22	1.95	2.45	3.09	3.90	4.92	6.20	7.83	9.87
210	0.07	0.11	0.18	0.29	0.46	0.73	1.16	1.86	2.33	2.95	3.71	4.68	5.91	7.46	9.40

Table 5.14 Wire lengths, in feet, for different AWG sizes of conductors with a specific ampacity and resistance for a 24 V system [46]. The first column contains the wire ampacity, in Ampere

AWG size	18	16	14	12	10	8	6	4	3	2	1	"1/0"	"2/0"	"3/0"	"4/0"
Copper uncoated DC resistance at 75 °C	7.95	4.99	3.14	1.98	1.24	0.78	0.49	0.31	0.25	0.19	0.15	0.12	0.10	0.08	0.06
1	30.19	48.10	76.43	121.21	193.55	308.48	488.80	779.22	979.59	1237.11	1558.44	1967.21	2481.90	3133.16	3947.37
2	15.09	24.05	38.22	60.61	96.77	154.24	244.40	389.61	489.80	618.56	779.22	983.61	1240.95	1566.58	1973.68
3	10.06	16.03	25.48	40.40	64.52	102.83	162.93	259.74	326.53	412.37	519.48	655.74	827.30	1044.39	1315.79
4	7.55	12.02	19.11	30.30	48.39	77.12	122.20	194.81	244.90	309.28	389.61	491.80	620.48	783.29	986.84
5	6.04	9.62	15.29	24.24	38.71	61.70	97.76	155.84	195.92	247.42	311.69	393.44	496.38	626.63	789.47
6	5.03	8.02	12.74	20.20	32.26	51.41	81.47	129.87	163.27	206.19	259.74	327.87	413.65	522.19	657.89
7	4.31	6.87	10.92	17.32	27.65	44.07	69.83	111.32	139.94	176.73	222.63	281.03	354.56	447.59	563.91
8	3.77	6.01	9.55	15.15	24.19	38.56	61.10	97.40	122.45	154.64	194.81	245.90	310.24	391.64	493.42
9	3.35	5.34	8.49	13.47	21.51	34.28	54.31	86.58	108.84	137.46	173.16	218.58	275.77	348.13	438.60
10	3.02	4.81	7.64	12.12	19.35	30.85	48.88	77.92	97.96	123.71	155.84	196.72	248.19	313.32	394.74
12	2.52	4.01	6.37	10.10	16.13	25.71	40.73	64.94	81.63	103.09	129.87	163.93	206.83	261.10	328.95
15	2.01	3.21	5.10	8.08	12.90	20.57	32.59	51.95	65.31	82.47	103.90	131.15	165.46	208.88	263.16
18	1.68	2.67	4.25	6.73	10.75	17.14	27.16	43.29	54.42	68.73	86.58	109.29	137.88	174.06	219.30
20	1.51	2.40	3.82	6.06	9.68	15.42	24.44	38.96	48.98	61.86	77.92	98.36	124.10	156.66	197.37
25	1.21	1.92	3.06	4.85	7.74	12.34	19.55	31.17	39.18	49.48	62.34	78.69	99.28	125.33	157.89
30	1.01	1.60	2.55	4.04	6.45	10.28	16.29	25.97	32.65	41.24	51.95	65.57	82.73	104.44	131.58
35	0.86	1.37	2.18	3.46	5.53	8.81	13.97	22.26	27.99	35.35	44.53	56.21	70.91	89.52	112.78
40	0.75	1.20	1.91	3.03	4.84	7.71	12.22	19.48	24.49	30.93	38.96	49.18	62.05	78.33	98.68
45	0.67	1.07	1.70	2.69	4.30	6.86	10.86	17.32	21.77	27.49	34.63	43.72	55.15	69.63	87.72
50	0.60	0.96	1.53	2.42	3.87	6.17	9.78	15.58	19.59	24.74	31.17	39.34	49.64	62.66	78.95

(continued)

Table 5.14 (continued)

AWG size	18	16	14	12	10	8	6	4	3	2	1	"1/0"	"2/0"	"3/0"	"4/0"
Copper uncoated DC resistance at 75 °C	7.95	4.99	3.14	1.98	1.24	0.78	0.49	0.31	0.25	0.19	0.15	0.12	0.10	0.08	0.06
55	0.55	0.87	1.39	2.20	3.52	5.61	8.89	14.17	17.81	22.49	28.34	35.77	45.13	56.97	71.77
60	0.50	0.80	1.27	2.02	3.23	5.14	8.15	12.99	16.33	20.62	25.97	32.79	41.37	52.22	65.79
65	0.46	0.74	1.18	1.86	2.98	4.75	7.52	11.99	15.07	19.03	23.98	30.26	38.18	48.20	60.73
70	0.43	0.69	1.09	1.73	2.76	4.41	6.98	11.13	13.99	17.67	22.26	28.10	35.46	44.76	56.39
75	0.40	0.64	1.02	1.62	2.58	4.11	6.52	10.39	13.06	16.49	20.78	26.23	33.09	41.78	52.63
80	0.38	0.60	0.96	1.52	2.42	3.86	6.11	9.74	12.24	15.46	19.48	24.59	31.02	39.16	49.34
85	0.36	0.57	0.90	1.43	2.28	3.63	5.75	9.17	11.52	14.55	18.33	23.14	29.20	36.86	46.44
90	0.34	0.53	0.85	1.35	2.15	3.43	5.43	8.66	10.88	13.75	17.32	21.86	27.58	34.81	43.86
95	0.32	0.51	0.80	1.28	2.04	3.25	5.15	8.20	10.31	13.02	16.40	20.71	26.13	32.98	41.55
100	0.30	0.48	0.76	1.21	1.94	3.08	4.89	7.79	9.80	12.37	15.58	19.67	24.82	31.33	39.47
110	0.27	0.44	0.69	1.10	1.76	2.80	4.44	7.08	8.91	11.25	14.17	17.88	22.56	28.48	35.89
120	0.25	0.40	0.64	1.01	1.61	2.57	4.07	6.49	8.16	10.31	12.99	16.39	20.68	26.11	32.89
130	0.23	0.37	0.59	0.93	1.49	2.37	3.76	5.99	7.54	9.52	11.99	15.13	19.09	24.10	30.36
140	0.22	0.34	0.55	0.87	1.38	2.20	3.49	5.57	7.00	8.84	11.13	14.05	17.73	22.38	28.20
150	0.20	0.32	0.51	0.81	1.29	2.06	3.26	5.19	6.53	8.25	10.39	13.11	16.55	20.89	26.32
160	0.19	0.30	0.48	0.76	1.21	1.93	3.05	4.87	6.12	7.73	9.74	12.30	15.51	19.58	24.67
170	0.18	0.28	0.45	0.71	1.14	1.81	2.88	4.58	5.76	7.28	9.17	11.57	14.60	18.43	23.22
180	0.17	0.27	0.42	0.67	1.08	1.71	2.72	4.33	5.44	6.87	8.66	10.93	13.79	17.41	21.93
190	0.16	0.25	0.40	0.64	1.02	1.62	2.57	4.10	5.16	6.51	8.20	10.35	13.06	16.49	20.78
200	0.15	0.24	0.38	0.61	0.97	1.54	2.44	3.90	4.90	6.19	7.79	9.84	12.41	15.67	19.74
210	0.14	0.23	0.36	0.58	0.92	1.47	2.33	3.71	4.66	5.89	7.42	9.37	11.82	14.92	18.80

Table 5.15 Wire lengths, in feet, for different AWG sizes of conductors with a specific ampacity and resistance for a 48 V system [46]. The first column contains the wire ampacity, in Ampere

AWG Size	18	16	14	12	10	8	6	4	3	2	1	"1/0"	"2/0"	"3/0"	"4/0"
Copper uncoated DC resistance at 75°C	7.95	4.99	3.14	1.98	1.24	0.78	0.49	0.31	0.25	0.19	0.15	0.12	0.10	0.08	0.06
1	60.38	96.19	152.87	242.42	387.10	616.97	977.60	1558.44	1959.18	2474.23	3116.88	3934.43	4963.81	6266.32	7894.74
2	30.19	48.10	76.43	121.21	193.55	308.48	488.80	779.22	979.59	1237.11	1558.44	1967.21	2481.90	3133.16	3947.37
3	20.13	32.06	50.96	80.81	129.03	205.66	325.87	519.48	653.06	824.74	1038.96	1311.48	1654.60	2088.77	2631.58
4	15.09	24.05	38.22	60.61	96.77	154.24	244.40	389.61	489.80	618.56	779.22	983.61	1240.95	1566.58	1973.68
5	12.08	19.24	30.57	48.48	77.42	123.39	195.52	311.69	391.84	494.85	623.38	786.89	992.76	1253.26	1578.95
6	10.06	16.03	25.48	40.40	64.52	102.83	162.93	259.74	326.53	412.37	519.48	655.74	827.30	1044.39	1315.79
7	8.63	13.74	21.84	34.63	55.30	88.14	139.66	222.63	279.88	353.46	445.27	562.06	709.12	895.19	1127.82
8	7.55	12.02	19.11	30.30	48.39	77.12	122.20	194.81	244.90	309.28	389.61	491.80	620.48	783.29	986.84
9	6.71	10.69	16.99	26.94	43.01	68.55	108.62	173.16	217.69	274.91	346.32	437.16	551.53	696.26	877.19
10	6.04	9.62	15.29	24.24	38.71	61.70	97.76	155.84	195.92	247.42	311.69	393.44	496.38	626.63	789.47
12	5.03	8.02	12.74	20.20	32.26	51.41	81.47	129.87	163.27	206.19	259.74	327.87	413.65	522.19	657.89
15	4.03	6.41	10.19	16.16	25.81	41.13	65.17	103.90	130.61	164.95	207.79	262.30	330.92	417.75	526.32
18	3.35	5.34	8.49	13.47	21.51	34.28	54.31	86.58	108.84	137.46	173.16	218.58	275.77	348.13	438.60
20	3.02	4.81	7.64	12.12	19.35	30.85	48.88	77.92	97.96	123.71	155.84	196.72	248.19	313.32	394.74
25	2.42	3.85	6.11	9.70	15.48	24.68	39.10	62.34	78.37	98.97	124.68	157.38	198.55	250.65	315.79
30	2.01	3.21	5.10	8.08	12.90	20.57	32.59	51.95	65.31	82.47	103.90	131.15	165.46	208.88	263.16
35	1.73	2.75	4.37	6.93	11.06	17.63	27.93	44.53	55.98	70.69	89.05	112.41	141.82	179.04	225.56
40	1.51	2.40	3.82	6.06	9.68	15.42	24.44	38.96	48.98	61.86	77.92	98.36	124.10	156.66	197.37
45	1.34	2.14	3.40	5.39	8.60	13.71	21.72	34.63	43.54	54.98	69.26	87.43	110.31	139.25	175.44
50	1.21	1.92	3.06	4.85	7.74	12.34	19.55	31.17	39.18	49.48	62.34	78.69	99.28	125.33	157.89

(continued)

Table 5.15 (continued)

AWG Size	18	16	14	12	10	8	6	4	3	2	1	"1/0"	"2/0"	"3/0"	"4/0"
Copper uncoated DC resistance at 75°C	7.95	4.99	3.14	1.98	1.24	0.78	0.49	0.31	0.25	0.19	0.15	0.12	0.10	0.08	0.06
55	1.10	1.75	2.78	4.41	7.04	11.22	17.77	28.34	35.62	44.99	56.67	71.54	90.25	113.93	143.54
60	1.01	1.60	2.55	4.04	6.45	10.28	16.29	25.97	32.65	41.24	51.95	65.57	82.73	104.44	131.58
65	0.93	1.48	2.35	3.73	5.96	9.49	15.04	23.98	30.14	38.07	47.95	60.53	76.37	96.40	121.46
70	0.86	1.37	2.18	3.46	5.53	8.81	13.97	22.26	27.99	35.35	44.53	56.21	70.91	89.52	112.78
75	0.81	1.28	2.04	3.23	5.16	8.23	13.03	20.78	26.12	32.99	41.56	52.46	66.18	83.55	105.26
80	0.75	1.20	1.91	3.03	4.84	7.71	12.22	19.48	24.49	30.93	38.96	49.18	62.05	78.33	98.68
85	0.71	1.13	1.80	2.85	4.55	7.26	11.50	18.33	23.05	29.11	36.67	46.29	58.40	73.72	92.88
90	0.67	1.07	1.70	2.69	4.30	6.86	10.86	17.32	21.77	27.49	34.63	43.72	55.15	69.63	87.72
95	0.64	1.01	1.61	2.55	4.07	6.49	10.29	16.40	20.62	26.04	32.81	41.42	52.25	65.96	83.10
100	0.60	0.96	1.53	2.42	3.87	6.17	9.78	15.58	19.59	24.74	31.17	39.34	49.64	62.66	78.95
110	0.55	0.87	1.39	2.20	3.52	5.61	8.89	14.17	17.81	22.49	28.34	35.77	45.13	56.97	71.77
120	0.50	0.80	1.27	2.02	3.23	5.14	8.15	12.99	16.33	20.62	25.97	32.79	41.37	52.22	65.79
130	0.46	0.74	1.18	1.86	2.98	4.75	7.52	11.99	15.07	19.03	23.98	30.26	38.18	48.20	60.73
140	0.43	0.69	1.09	1.73	2.76	4.41	6.98	11.13	13.99	17.67	22.26	28.10	35.46	44.76	56.39
150	0.40	0.64	1.02	1.62	2.58	4.11	6.52	10.39	13.06	16.49	20.78	26.23	33.09	41.78	52.63
160	0.38	0.60	0.96	1.52	2.42	3.86	6.11	9.74	12.24	15.46	19.48	24.59	31.02	39.16	49.34
170	0.36	0.57	0.90	1.43	2.28	3.63	5.75	9.17	11.52	14.55	18.33	23.14	29.20	36.86	46.44
180	0.34	0.53	0.85	1.35	2.15	3.43	5.43	8.66	10.88	13.75	17.32	21.86	27.58	34.81	43.86
190	0.32	0.51	0.80	1.28	2.04	3.25	5.15	8.20	10.31	13.02	16.40	20.71	26.13	32.98	41.55
200	0.30	0.48	0.76	1.21	1.94	3.08	4.89	7.79	9.80	12.37	15.58	19.67	24.82	31.33	39.47
210	0.29	0.46	0.73	1.15	1.84	2.94	4.66	7.42	9.33	11.78	14.84	18.74	23.64	29.84	37.59

Wire Between the Inverter and the AC Distribution Panel
The highest continuous current that might be drained from the batteries by the inverter is 12.5 A. The cable should not be loaded with more than 80% current. Therefore, the current rating of the wire should be at least:

$$12.5\,A \times 125\% = 15.625\,A.$$

From Table 310.16 (NEC 2020), the wire of choice is THHW (75 °C) type #14 AWG copper wire, which can carry up to 20 A of current. The accepted voltage drop in the AC side is not more than 1.5%. Therefore, the voltage drop should not exceed 240 V $\times 1.5\% = 3.6$ V for this wire.

The inverter is 30 ft away from the AC distribution panel. So, the wire has to be taken for twice this distance, i.e., $2 \times 30 = 60\,\text{ft} = 0.06\,\text{kft}$. According to Table 9, Chapter 9 of NEC 2020, the AC resistance of the #14 AWG wire size is 2.7 ohm/kft. Therefore, the voltage drop in the #14 AWG copper wire is

$$V_{drop} = 12.5\,A \times 2.7\,ohm/kft \times 0.06\,kft = 2.025\,V,$$

which is less than 3.6 V. Therefore, the selected wire on the AC side is #14 AWG copper wire, which can safely carry up to 20 A of current.

5.6.2 Overcurrent Protection Devices

In addition to all safety factors discussed in the wiring sizing section, the wires or conductors should still be protected against any current greater than the rated allowable current. An overheating problem occurs when an overcurrent passes through the conductor, which may eventually cause a fire. Therefore, overcurrent protection devices (OCPD) are highly recommended to ensure good operating conditions for the conductors and avoid any possible fire cause. The OCPD can be classified based on their timely response to circuit breakers and fuses, which are the most used devices in AC and DC sides of the PV system. Both have a nominal ampacity rating after which the fuse blows, or the circuit breaker trips to stop the current flow by opening the electrical circuit. The response time varies based on the technology used in the device. In contrast, the fuse is mainly made of a metal wire or a filament placed in a metallic, glass, or ceramic casing. When the power flow exceeds the nominal rated power, the metal wire starts to heat up until it melts, so it physically opens the electrical circuit. The fuses are inexpensive and always available. However, the response time is relatively slow. In addition, it cannot be used as a disconnect and OCPD because it cannot be removed or replaced under load, unlike circuit breakers. On the other hand, circuit breakers are based on electromagnetic technology. The overcurrent generates a magnetic force that compels metallic contact to change its position and open the circuit. These devices are more expensive than fuses. However, they respond faster and can also be used as disconnects when the system needs maintenance, or any part of the system needs to be replaced. The OCPD is required in the DC side at the output of a PV source circuit and at the output of the inverter side where AC power is generated.

5.6.2.1 DC Side

Series fuses are usually used on the output of each PV source circuit as OCPD. The outputs of the PV source circuit pass through the fuses and get connected to the common point, the combiner box. The ampacity rate of the fuse should be not less than 156% of the PV source circuit. For example, the PV power source consists of 5 modules connected in series, whereas each module has a short circuit of 8 A.

The fuse ampere rating is $8 \times 156\% = 12.48$ A. The minimum accepted ampere rating is 12.48 A. Therefore, by checking Table 240.6(A) of NEC 2020, the nearest round-up value available in the market is 15 A. It is the proper fuse of choice.

According to Article 690.9(A) of NEC 2020 [45], there are two conditions where the series fuses are not required:

1. There are no external source circuits, such as batteries, parallel source circuits, or backfeeding current from inverters. This case occurs when the system has only one PV power circuit and an inverter. The reason behind this exception is that the PV source circuit has a limited current output. So, it will not produce more than 125% of its short-circuit current in the worst-case scenario. Therefore, this fault current value (125% of the short-circuit current) will not exceed the fuse current rating, which is supposed to handle 156% of the short-circuit current. So, the fuse will not operate in this condition.
2. The maximum short-circuit current of all the sources does not exceed the ampacity of the conductors and the OCPD of the PV module or the inverter. Suppose a system has 2 PV source circuits. When one of these circuits is under a short-circuit fault, the current provided by the other circuit will start feeding the faulty part of the shorted circuit. This is known as backfeeding—when a fault causes the current to flow back toward the PV modules. However, the value of this backfeeding current does not exceed 125% of the short-circuit current. So, a fuse will not be necessary in this case because the conductor is already rated for this increased current, i.e., it is within the withstand capacity of the wire, which is the maximum short-circuit current that the wire can withstand without damage.

The backfeeding process for three PV source circuits is shown in Fig. 5.37. Here, there is a short-circuit fault between two parts of the third source circuit. Due to the presence of this low resistive path, the currents from the other two circuits start to flow back toward the faulty point. This current may be up to $(2 \times 1.25 \times I_{SC})$ and enough to damage the PV modules and other equipment in the circuit. This is a dangerous situation in a solar PV system that requires the use of series fuses on the DC side for protection.

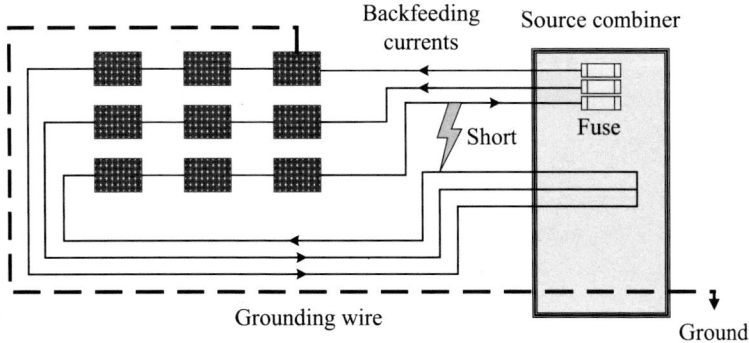

Fig. 5.37 Three PV source circuits backfeeding a short-circuit PV source circuit [47]

5.6.2.2 AC Side

The OCPD is also installed on the inverter's output side. The minimum current rating of OCPD is 125% of the nominal current of the inverter as provided by the manufacturers' datasheet.

Disconnects

In order to fulfill the safety replacement and maintenance requirements, disconnects are highly recommended by the NEC for both grounded or ungrounded PV systems on both the AC and DC sides. Disconnects are required for all ungrounded conductors to isolate them from the other system components. Moreover, they prevent any possible electric shock or fault current that might cause fatal accidents for the maintenance electrician.

Most PV system designers ground the negative DC terminals and the neutral terminal on the AC side. So it is not essential to apply disconnects on these terminals, unlike ungrounded positive terminals, which can be dangerous. The disconnects are not only limited to PV source circuits and inverters, but they can also be applied to batteries and charge controllers. It is worth noting that AC and DC side disconnects are now embedded in all residential and commercial grid-tied inverters. These disconnects are designed to handle 1.25 times the maximum continuous current of the inverter. PV system disconnect arrangement is given in Fig. 5.38.

5.6.3 Switches

Typically, a solar PV system comes with two safety switches or disconnects. The first one is the DC disconnect/switch, which can interrupt the flow of the DC current between the solar module (source) and the inverter by opening the circuit. In some cases, it is integrated into the inverter. The second one is called AC

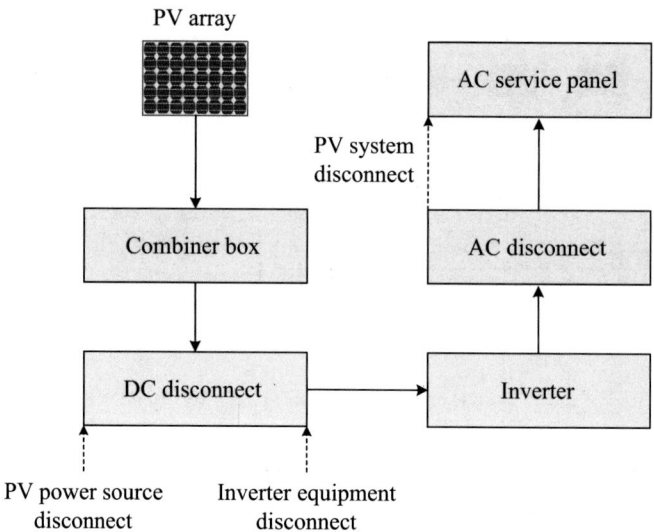

Fig. 5.38 PV system disconnects [48]

disconnect/switch, which separates the inverter from the load. Normally, they are installed after the inverter and near the consumer unit (on exterior walls when used at home).

The switching devices are developed for a reliable operation in a wide variety of solar PV applications in residential buildings and industrial grounds. In cases of disasters such as a fire in or around a home, a user can shut the inflow of power by using DC and AC disconnects that can lower the spread of fire and prevent the risk of electrocution. During natural calamities, disconnecting the system at the DC disconnect will protect the inverter and interior wiring of the house.

5.6.3.1 Temperature Effect

Switches, along with other equipment in a solar PV system, have to deal with a wide range of temperatures from −40 °C to +55 °C or even higher. The thermal current rating of a switch device carries the maximum current value that is allowed to pass through without heating it up beyond the tolerable range. An allowable temperature rise of 70 °C under normal circumstances implies that in a 24 h period, the accumulated heat of environmental temperature adds up to the 70 °C. Suppose the ambient temperature of 35 °C adds with the 70 °C temperature rise, resulting in a combined total of 105 °C, which is the highest temperature permitted in a switching device. But in a PV system, the ambient temperature can go beyond normal atmospheric temperatures. The higher the ambient temperature, the lower the allowed temperature rise. Because in higher temperatures, a small current can heat up the switch up to the highest permitted temperature. On the contrary, colder temperatures do not pose any adverse impacts on the switch.

Table 5.16 Derating factor (DF) for solar PV disconnect at different ambient temperatures [50]

Disconnect rating	Temperature				
	15 °C [59 °F]	20 °C [68 °F]	40 °C [104 °F]	55 °C [131 °F]	70 °C [158 °F]
15 A	18 A	17 A	15 A	13 A	11 A
DF	1.18	1.15	1	0.87	0.72
25 A	28 A	28 A	25 A	25 A	20 A
DF	1.13	1.11	1	0.91	0.81
40 A	45 A	44 A	40 A	37 A	33 A
DF	1.12	1.10	1	0.92	0.83
70 A	80 A	79 A	70 A	62 A	52 A
DF	1.15	1.13	1	0.98	0.75

Temperature derating is the controlled reduction of power to prevent overheating issues due to the excessive flow of current. The calculation of the derating factor is based on the maximum allowed temperature rise values. If temperature derating is necessary, an appropriate calculation can be utilized to obtain estimated values for the temperature derating factors [49].

$$DF = \sqrt{\frac{T_{max} + \Delta_{max} - T_{ambient}}{\Delta_{max}}}, \tag{5.27}$$

where

T_{max} is the maximum allowed average temperature under normal situation,
$T_{ambient}$ is the ambient temperature, and
Δ_{max} is the maximum allowed temperature rise (70 °C).

So, for an ambient temperature of 60 °C in a switch, the derating factor would be

$$DF = \sqrt{\frac{35 + 70 - 60}{70}} = 0.8. \tag{5.28}$$

Table 5.16 describes how the rating of disconnect varies with different temperatures. Temperature factors of the respective temperature are expressed inside square brackets. From the table, it can be seen that, when the ambient temperature is greater than or less than 40 °C, the current rating is modified accordingly by dividing by the DF at that temperature.

5.6.3.2 Switch Sizing

Switches or disconnects come in different sizes and are used accordingly as per requirement. According to the system specifications, a prudent calculation is necessary to decide the appropriate size of the switches for the system. The parameters that are considered to determine the overall load size are circuit load, voltage, amps/breaker size, and the wiring and cable size. Determining the overall load size will determine the required switch size.

Example 5.10 (DC Disconnect Sizing) Suppose a solar PV array contains 20 parallel-connected strings, and each of those strings contains 30 modules connected in series. Each of those modules has a V_{OC} of 28.4 V and I_{SC} of 7.92 A. The V_{OC} of the PV array determines the voltage rating for the DC disconnect. Here, the total voltage of the PV array is

$$V_{OC} = 30 \times 28.4\ V = 852\ V.$$

The sum of the I_{SC} of the parallel-connected strings is

$$I_{SC} = 20 \times 7.92\ A = 158.4\ A.$$

According to Article 690.8(A)(1) of NEC 2020 [45], the maximum current of a PV system circuit will be 1.25 times the sum of the I_{SC} of the parallel-connected modules. So, considering a 1.25 higher rating, the current rating would be

$$I_{SC} = 158.4 \times 1.25 \times 1.25 = 247.104\ A.$$

In case of the rise of ambient temperature around the installation site (e.g., up to 60 °C), a temperature derating factor needs to be considered. For a temperature of 60 °C, the derating factor is 0.8, according to Eq. 5.28. So, under normal conditions, I_{SC} would be

$$I_{SC} = \frac{247.104}{0.8} = 308.88\ A.$$

The DC disconnect should be rated at the total open-circuit voltage and the total short-circuit current of the PV array, as per Article 690.13(E) of NEC 2020. Since the voltage and the current in this example are 852 V and 308.88 A, respectively, so using a 1000 V, 400 A ZJBENY BH-400 DC disconnect [51] will suffice to isolate the system from the inverter. Some common sizes of DC disconnects are 32 A 250 V, 20 A 450 V, 16 A 500 V, 250 A 500 V, 400 A 500 V, 250 A, and 1000 V.

Example 5.11 (AC Disconnect Sizing) An AC disconnect is installed between the home's main service panel and inverter. This disconnect receives the output current of the inverter. So, the maximum output current listed on the inverter datasheet should be considered. The current rating for an AC disconnect would be 1.25 times the inverter AC output current. Let us consider the SMA Sunny Boy 3.6 kW inverter [52] that has an output current rating of 16 A. So, the required ampacity of the AC disconnect will be

$$16 \times 1.25 = 20\ A.$$

The disconnect rating suitable for the mentioned inverter should be at least 20 A. From the available disconnects in the market, the Siemens 30 A WF2030 Fusible AC disconnect [53] would be a suitable choice for this system, since it can carry up to 30 A of current. To learn more about switch sizing, visit Refs. [54, 55].

5.6.4 Grounding

In electrical engineering, the ground refers to the earth, which is regarded to have no electric potential, i.e., a potential of 0 V. Grounding means connecting the negative terminal of a DC system or the neutral conductor of an AC system to the ground. This creates a low resistive path against lightning or any other faults. Grounding can be done in two ways: solid grounding without any resistive component or grounding through a resistor. In continuation of the safety requirements that the designers should meet, NEC mandates that PV systems should satisfy system and types of equipment grounding standards. These standards ensure good performance and safety for the personnel and the system components. Moreover, it protects from any possible ground fault current or lighting surge that is a possible reason for electrocution and electric shocks.

5.6.4.1 Importance of Grounding
Grounding provides an alternative current path to all harmful, high currents resulted from undesirable conditions, such as lighting, electrical faults, phase unbalance, or any short circuit between the metallic current-carrier contact and the metallic frame of components, such as the inverter case or the frame of the PV module. These currents may cause electrocution and voltage shocks when passing through human bodies when grounding does not exist, as shown in Fig. 5.39. Moreover, grounding plays a role in stabilizing the voltage while providing a standard reference to all voltages with the ground that has a zero-volt potential. So, all electrical systems must be securely grounded in order to protect both life and equipment.

5.6.4.2 Types of Grounding
Electrical grounding is mainly of two types: system grounding and equipment grounding:

1. **System Grounding:** System grounding refers to the grounding of all negative current-carrying conductor terminals of the PV source circuit, arrays, and batteries. So, the positive current-carrier conductors are the ungrounded conductors. The AC side should also be grounded by shorting the neutral conductors to the ground in the main distribution panel. It is worth noting that in some cases, such as in batteries, the DC positive terminal is the grounded terminal. Consider a 12 V battery as an example. Grounding the positive terminal makes its potential 0 V, which was 12 V before grounding. But the potential difference between the two terminals of a battery should always be constant; so, the negative terminal

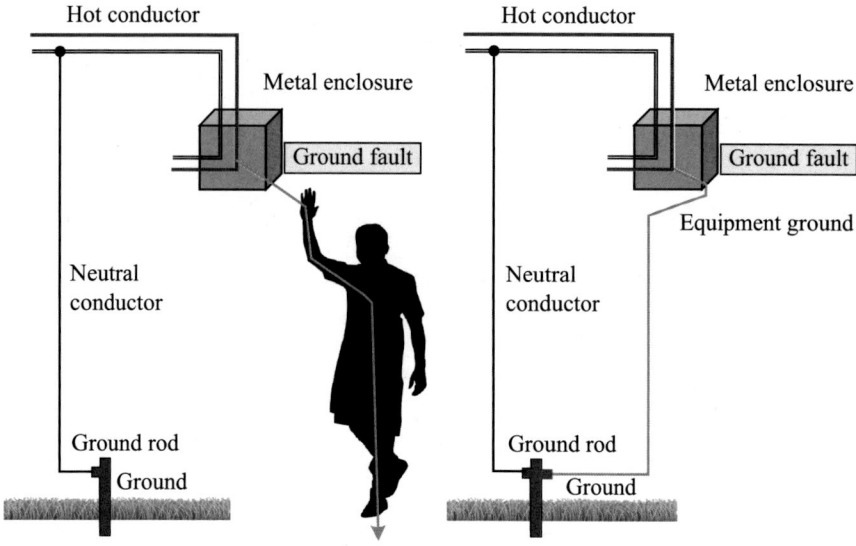

Fig. 5.39 The importance of grounding systems [48]

of the battery becomes -12 V. Thus, a negative voltage can be obtained from a battery. In contrast, the phase wire is never grounded in AC systems.

2. **Equipment Grounding:** If an ungrounded naked conductor is connected to the metallic chassis of electrical equipment, the chassis is electrified and is a source of danger when exposed to human touch. When a human touches this energized metallic chassis, the circuit is closed using the human body by providing a low resistive path to the current. This can cause burn injuries and even death based on the current flow through the body. Therefore, a very low resistive conductor should be shorted to the metallic chassis of all components and devices to provide an easier path to the fault current than the human body. This is called equipment grounding. This conductor does not contain any current or voltage in the normal operating conditions unless a fault or unbalance occurs. The Aluminum frames of solar modules should be grounded with the mounting system where the grounding cable goes down to a wet soil that should be frequently maintained to ensure wetness and resistivity.

The grounding wire sizing process depends on the OCPD sizes, as illustrated in Table 250.122 provided in NEC 2020. For example, when a 100 A OCPD device is selected, the grounding conductor should not be less than #8 AWG if it is copper and #6 AWG if it is aluminum or copper-clad aluminum. In addition, these grounding conductors are usually colored yellow and green.

5.6.4.3 Step and Touch Potential

The resistance of the human body ranges from 1 to $50\,k\Omega$ in dry conditions. Despite the resistance, the human body is a conductor of electricity due to the presence of various salts in our bloodstream in an ionic form. So, electric currents often end up flowing through the body, causing long-term damages to the body, and even death. Two concerning terminologies in this regard are the step potential and the touch potential.

The ground surrounding high voltage electrical equipment may get charged. In that case, if a person stands within that area, then a potential difference is created between the feet of the person. This potential is known as the step potential. If such a potential exists, current can flow through the person's body from one foot to the other.

If a person touches an electrically charged equipment, then a potential difference arises between the person's hand and the ground. This potential is known as the touch potential. If such a potential exists, current can flow through the person's body.

Both step and touch potentials are dangerous conditions, especially during a ground fault, and may even claim the life of the person. These two conditions can be avoided by appropriate grounding through a grounding conductor and by ensuring insulating footwear and insulating hand gloves for the person to break the path of current.

5.6.4.4 Some Definitions Related to Grounding [45]

1. Ground fault: A ground fault, or earth fault, is a type of short-circuit fault in which the deviating current flows directly to the ground. Such faults result in high circulating currents, charging up of the non-current-carrying parts of the system, and increase the risk of electrocution.

2. Grounding electrode conductor (GEC): The GEC is a wire or busbar type conductor made of copper, aluminum, or copper-clad aluminum. It connects the equipment or system grounded conductor to the ground rod or the grounding electrode. For sizing the GEC of a system, Table 250.66 of NEC 2020 [45] has to be followed.

3. Equipment grounding conductor (EGC): The EGC is a conductor made of copper, aluminum, or copper-clad aluminum that connects the non-current-carrying conducting parts of an electrical equipment to the ground rod, the GEC, or both. For instance, the metallic parts of equipment can be energized. The EGC can send this current to the ground, thus ensuring the safety of personnel who may touch the equipment. For sizing the EGC of a system, Table 250.122 of NEC 2020 [45] has to be followed.

4. Ground fault circuit interrupter (GFCI): According to Article 100 of NEC 2020, a GFCI is a device meant for personnel protection. It deenergizes a circuit or a part of it within a specific period of time if the line to ground current exceeds the rated value of the circuit.

The minimum length or depth of the grounding electrode should be 8 ft, according to Article 250.52 of NEC 2020 [45]. In addition, two grounding electrodes must have a minimum distance of 6 ft between them. Instead of inserting multiple grounding electrodes inside the ground, it is also possible to connect all the neutral or grounding wires in one bus bar and then ground the bus bar using one grounding electrode.

5.6.5   Lightning Protection

Since PV arrays can be installed on elevated roofs, they might be exposed to lightning strikes. When a lightning strike hits the metallic mounting system, it needs to find a low resistance path to the ground and might pass through the wires to reach the inverter or battery system. This current is usually in the kA range and is capable of damaging everything in the process. To avoid this situation, the metallic aluminum frame of the modules and the mounting system should be grounded. Grounding provides an alternative path to the high current that needs to be dissipated in the wet ground. Such a grounding mechanism can be achieved by using a lightning rod, which is a low resistive conductor placed at a higher elevation than the device to be protected, and directly connected to the ground so that the rod receives the lighting before any other device touches it, and conducts the voltage straight to the ground. The importance of lightning protection can be perceived from Fig. 5.40. The connection of the lightning rod can be observed in Fig. 5.41.

To ensure safety from lightning strikes, a lightning arrestor or a surge arrestor can be used. Both can be used for lightning protection; however, a lightning arrestor cannot protect against a surge, which is a high-magnitude transient state of voltage, current, or power in an electric circuit. Surges can damage the whole system if not defended. Article 242.42 of NEC 2020 specifies the selection of surge arrestors. Its rating should not be less than the maximum continuous operating voltage at the point of application. In case of a solidly grounded system, this voltage is the phase to ground voltage of the system. But for a system grounded through an impedance or an ungrounded system, this voltage is the phase to phase voltage of the system. For a silicon carbide type surge arrestor, the rating should be multiplied by a further 125%.

5.6.6   Net Metering

Any user having a solar PV system installed may decide to use energy either from the utility grid or the solar PV system. Consuming energy from the grid may be costly at peak hours. At these times, it is better to use energy from the PV system. If the output of the PV system is more than the required amount, the excess energy can be stored in a battery. A hybrid PV system allows a user to send surplus PV energy into the utility grid after proper synchronization. So, the user can earn credits by selling PV energy to the grid, which can help to cut down on the cost of consuming

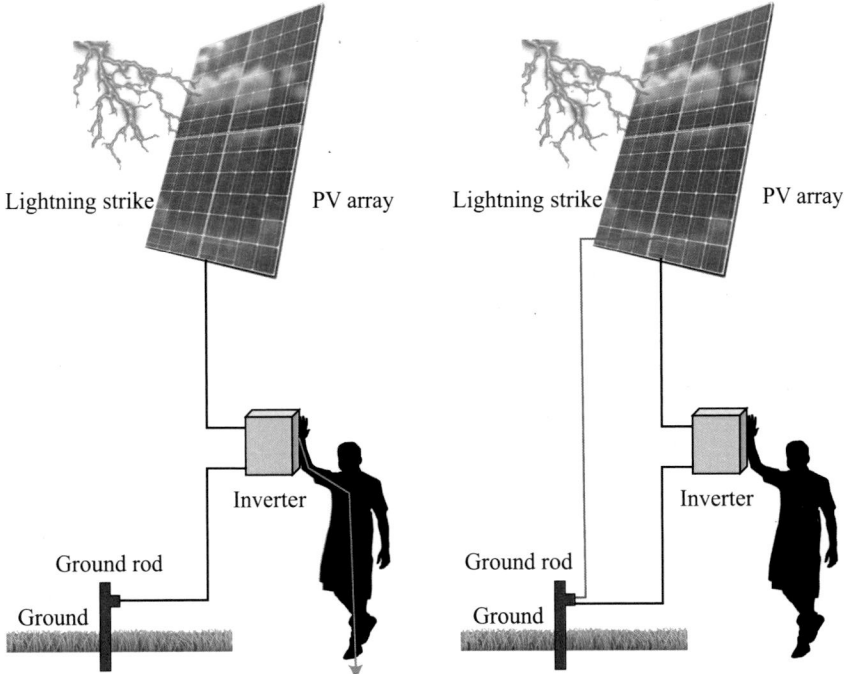

Poor grounding = Poor lightning protection   Good grounding = Better lightning protection

Fig. 5.40 The importance of lightning protection systems [48]

energy from the grid. This billing system is referred to as a net metering system. This system contains a bidirectional energy meter that can record the net energy flow from the user to the grid. Figure 5.42 illustrates the basic theme of net metering.

5.6.7   Module Mounting

The capacity of solar PV power plants is growing day by day. PV mounting systems are used to attach solar modules on the ground, on rooftops, or building facades. The necessity of understanding the many aspects and types of mounting structures cannot be overstated. Solar PV modules have a life of 25–30 years. So, it is important to choose the right kind of solar PV mounting structure to support it for this long period of time. Typically, these module mounting structures are made of three types of materials—aluminum, hot-dip galvanized iron, and mild steel. Aluminum is a corrosion resistive material, and it is lightweight and cost-effective compared to galvanized iron. Galvanizing iron is prepared by applying galvanizing zinc coating to iron or steel to prevent it from corroding. Hot-dip galvanization is the most popular method of galvanizing that is employed in module

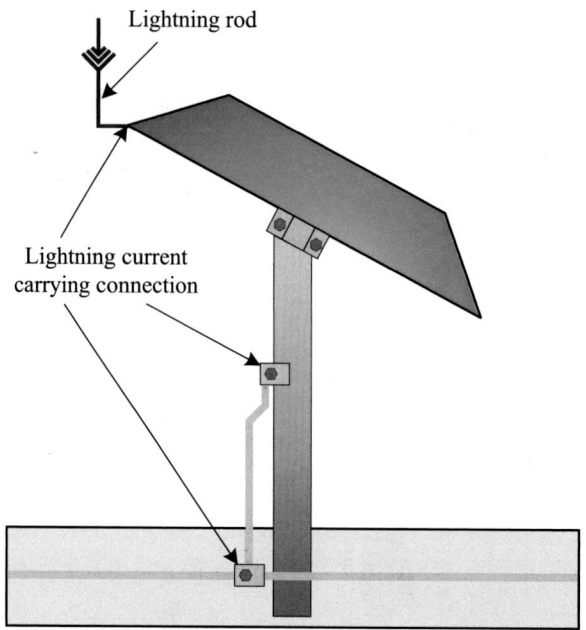

Fig. 5.41 The connection of the lightning rod

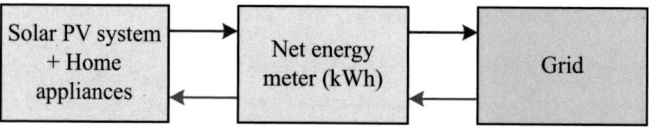

Fig. 5.42 Block diagram representing the process of net energy metering through a net meter [56]

mounting structures. Mild steel is not used very often to manufacture solar PV module mounting. It is used where lightweight is required, such as not so strong rooftops. It contains a very low amount of carbon and is highly flexible.

In terms of racking construction and mounting options, each PV system is unique, and its installation condition differs from one place to another. Therefore, many types of solutions have been developed to house different requirements such as pole mount, ground mount, and roof mount, which are demonstrated in Fig. 5.43.

5.6.7.1 Ground Mounted Solar Module

Ground mounted solar modules include solar modules that are mounted in the open place on the ground where they can receive a plenty of sunlight. Though rooftop solar modules are the most popular ones, ground solar modules are less complex to install. The modules feed power to an inverter that is installed either in the house or behind the mounting system. Typically, residential ground mounted

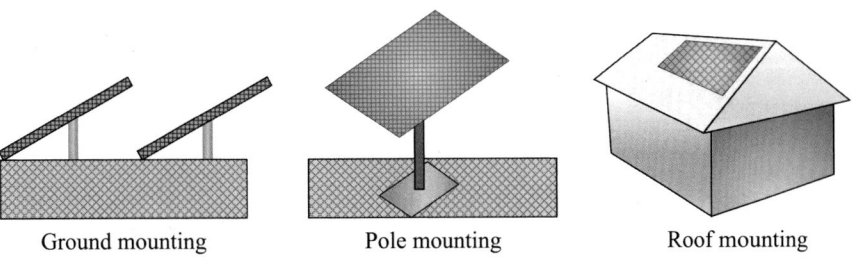

Ground mounting   Pole mounting   Roof mounting

Fig. 5.43 Three types of mounting for solar PV systems

solar installations are built using 60-cell solar modules, whereas in large-scale solar farms, a ground mounted solar PV system consists of 72-cell solar modules. However, bifacial or half-cell designs use 120 or 144 cells.

Advantages
1. It provides easy access to maintenance and troubleshooting.
2. It is easier to choose optimum tilt for maximum production for its flexible positioning.
3. Installation is safer as it does not require any roof climbing.

Disadvantages
1. It is more susceptible to vandalism, theft, and damages due to its easy access.
2. The biggest disadvantage is the large space requirement.
3. In case of a faulty design, PV modules can shade one another due to congested spacing or limited row–row spacing.

5.6.7.2 Pole-Mounted Solar Module
It is one of the most straightforward and trouble-free types of solar installation. Pole mounted systems are robust structures that use a single pole that can hold more than a dozen modules. It can be easily equipped with a Sun tracking system that can ensure maximum exposure. This arrangement is common in space-constrained public areas, such as parks or agricultural farms.

Advantages
1. Since the modules are elevated from the ground, they can achieve lower module temperatures with natural convection using air.
2. The positioning of these modules is also very flexible, and so, it helps to reach maximum production.
3. It leaves a small footprint as it typically uses one pole.

Disadvantages
1. It requires concrete, steel, trenching, and foundation for construction.
2. It is a costly solution.

Fig. 5.44 A 12.6 kW grid-tied rooftop solar PV system in Medford, Oregon, USA. This system employs a net metering system but has no battery backup

5.6.7.3 Roof Mounted Solar Module

These are the most popular types of solar modules in residential areas and commercial buildings. The types of roof mounted solar modules vary with the type of the roof it is installed on. When considering the design of a roof mounted solar PV system, the condition of the roof—whether the roof is fire resistant, waterproof, structurally strong, and meets electrical safety requirements or not—should be reviewed. Some of these roof mounted systems are flat roof mounting, sloped roof mounting, rail-less mounting, railed mounting, and shared-rail mounting system. A real rooftop solar PV system is illustrated in Fig. 5.44. Houses with a plain horizontal roof instead of an inclined one have to rely on a mounting system similar to a ground mounting system for setting up a solar PV system. Thus, the tilt angle of the module is limited by the roof structure.

Advantages
1. This structure is more cost-effective.
2. As it can be installed on rooftops, it saves a lot of space.
3. It is secure, as only the authorized individuals can have access.

Disadvantages
1. In some cases, the roof may require an additional support mechanism as the modules add additional weight to the roof.
2. Maintenance is difficult as it is not easily accessible. It can be a safety hazard for the installation team.
3. In case of installing this mounting system on shingle roofs, roof penetration is required.
4. Designers have no choice but to install the modules at the existing tilt angle of the roof instead of using the best tilt angle. However, this problem can be overcome if the roof tilt is originally designed keeping in mind the installation of solar modules.

5.6.8 Solar Tracking Systems

Since the Sun appears to move from the east to the west on a daily basis and also differs in its position and path in the sky on a seasonal basis, the maximum solar energy can be captured from the PV module if the module continuously faces toward the Sun. This idea has given birth to solar tracking mechanisms that automatically point toward the Sun in accordance with the perceived movement of the Sun in the sky. Tracking systems have been found to significantly raise the efficiency of solar PV modules [57]. Compared to fixed mounting, tracking systems make solar PV systems much more efficient (Fig. 5.45).

Based on the tracking axis, solar trackers may be of two types—single-axis tracker and dual-axis tracker. Figure 5.46 shows that the use of tracking systems

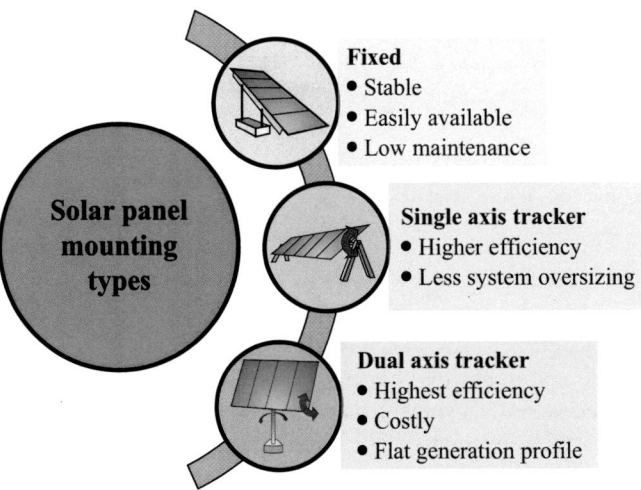

Fig. 5.45 The comparison between fixed mounted and single-axis and dual-axis tracking systems

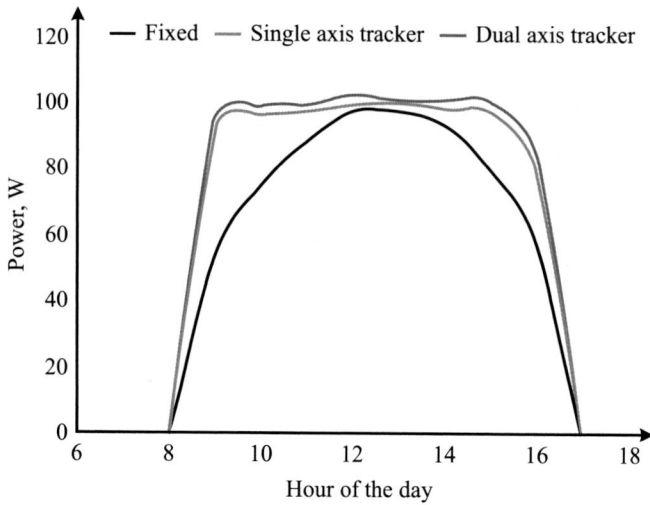

Fig. 5.46 The increase in the output power of solar PV modules due to the use of tracking systems

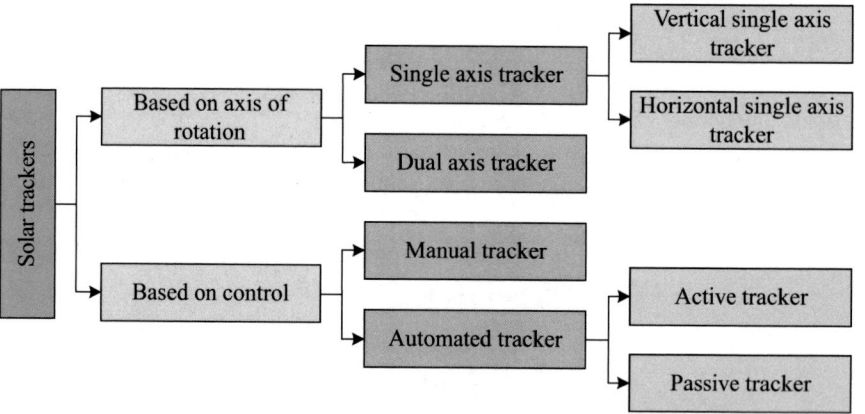

Fig. 5.47 Classification of solar tracking systems

result in an increase in the output power of the PV modules. Dual-axis trackers provide more output power more compared to single-axis trackers. Solar tracking systems are further classified as shown in Fig. 5.47. A single-axis tracker can rotate from east to west and tracks the Sun from sunrise till sunset. These trackers can again be classified as vertical single-axis tracker (VSAT) and horizontal single-axis tracker (HSAT), which are compared in Table 5.17 and Fig. 5.48.

If we observe the path of the Sun as an observer in the northern hemisphere, then it will be understood that after sunrise, the Sun not only moves upward in the sky (increasing altitude) toward the west, but also moves toward the south (increasing

Table 5.17 Comparative assessment of vertical single-axis tracker (VSAT) and horizontal single-axis tracker (HSAT) [58]

Characteristic	VSAT	HSAT
Axis orientation	Vertical	Horizontal
Location	Good at higher latitudes	Good at lower and mid latitudes
Land space requirement	More space required	Less space required
Power density per unit area	Lower power density	Higher power density

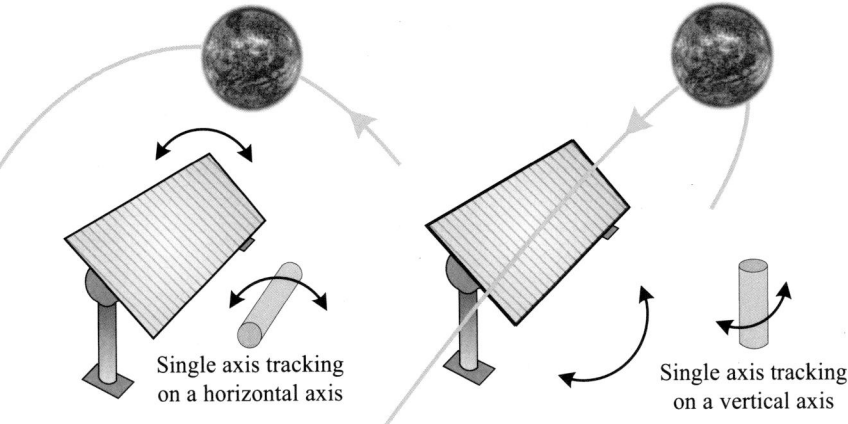

Single axis tracking
on a horizontal axis

Single axis tracking
on a vertical axis

Fig. 5.48 Two types of single-axis trackers—horizontal single-axis tracker (HSAT) and vertical single-axis tracker (VSAT)

azimuth angle). The dual component of the displacement of the Sun can be tracked using a dual-axis solar tracker. Besides the usual east–west tracking system, a dual-axis tracker has an additional tracking direction from north to south to track the path of the Sun. The north–south axis tracking mechanism of a dual-axis tracker helps the module to adjust itself with the changing altitudes of the Sun in different months of the year. The dual-axis rotation of a dual-axis tracker is illustrated in Fig. 5.49.

Although dual-axis solar trackers yield a better efficiency of the module and harvest more energy from the sunlight, they are much more expensive and employ a complicated mechanism compared to single-axis solar trackers. Due to the greater number of moving parts, the dual-axis trackers are also vulnerable to more mechanical wear and tear, which adds up to the maintenance and repair costs.

Based on the actuating mechanism, solar trackers may be manual or automatic. Manual trackers are moved by hand and involve employing dedicated personnel to move the solar module as the day advances, thus raising the operational cost. These systems are suitable for developing or underdeveloped countries with low employment rates and low labor costs. But for developed countries with high labor costs, it is wiser to deploy an automated tracking system. Automated systems are classified as passive and active systems. Passive tracking systems usually have a

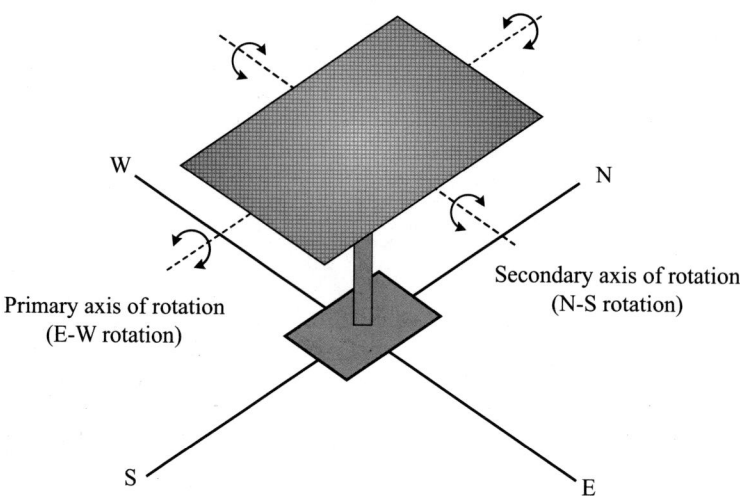

W

N

Secondary axis of rotation
(N-S rotation)

Primary axis of rotation
(E-W rotation)

S

E

Fig. 5.49 The two axes of rotation of a dual-axis solar tracker

liquid inside the rotating axis with a low boiling point. The liquid vaporizes using the heat from the sunlight and causes an imbalance within the axis, eventually making the module move. On the other hand, active tracking systems utilize electric motors or hydraulic rams to move the module.

5.6.9 PV Module Cooling

Since solar PV modules constantly face the Sun throughout the day, they are very likely to heat up. Similar to any other electrical device, PV modules may also heat up beyond their tolerable limits, leading to a reduction in their performance and efficiency. The efficiency reduces as the temperature rises, at the rate of nearly 0.5% reduction for each °C temperature rise. Figures 3.8 and 3.9 have already demonstrated that both the current and power outputs decline with the rise of the module temperature. Hence, an appropriate cooling system is essential for PV systems to maintain the temperature at an optimum level to maximize the PV module efficiency. It should be remembered that cooling is a prime necessity in the hot summer days but can be spared during the winter. During the rest of the year, cooling can be done whenever the temperature is too high, usually more than the standard temperature of the specific place. Hence, cooling may be termed as a seasonal necessity for the optimal operation of solar PV systems.

PV module cooling systems may be classified into two main types—active and passive, based on their power consumption. Active cooling systems require a power source to operate and are typically driven using a fan or a motor. On the other hand, passive cooling systems do not require any power source. Again, based on the fluid used for cooling, PV module cooling systems may be primarily divided into two types—air-based cooling and water-based cooling.

There are several types of cooling technologies for each type, but water-based cooling technologies have been found to be more effective than air-based ones [59]. Water has higher specific heat than that of air, implying that a lesser volume of water is able to absorb a greater amount of heat compared to the same volume of air. Since water is a limited and valuable resource, its use for cooling systems is not a wise choice given the present freshwater constraints. On the contrary, using saline water may create other physical and chemical conundrums for the PV modules. So, a sustainable choice is to use the freshwater for solar module cooling and then reusing the heated water for other purposes such as domestic heating, aquaculture, or other applications requiring warm water. Water cooling also has the added benefit of cleaning the PV modules, which eliminates the necessity of a separate cleaning scheme.

Water cooling systems typically have tiny water outlets in narrow pipes fitted at the top edge of the PV arrays that spray a thin jet of water that streams down along the array in a laminar flow, thereby absorbing the heat from the array upon contact. The water can be let to fall down into the ground, but a more efficient method is to catch the water at the lower end of the array in a gutter and recirculate it for reuse at a later time. The water flows through the gutter, is stored in a tank for later usage, and pumped again into the spray boom when cooling is required. Rainwater could also be reserved for this purpose. The system is automatically activated when a sensor detects a rise from the normal temperature of 25 °C. This technology has been developed by the French PV system installer Sunbooster, and it can boost the annual output of the PV system by 8–12% [60]. The schematic diagram of the system is shown in Fig. 5.50.

The automated cooling system described above comes with a major limitation— added costs. A solar PV system is a large investment in itself, and it is desired to cut down on unnecessary costs. So, most PV system owners prefer using a manual cooling system—using a regular hosepipe and washing down the modules with water or washing down the module with a damp brush or fabric.

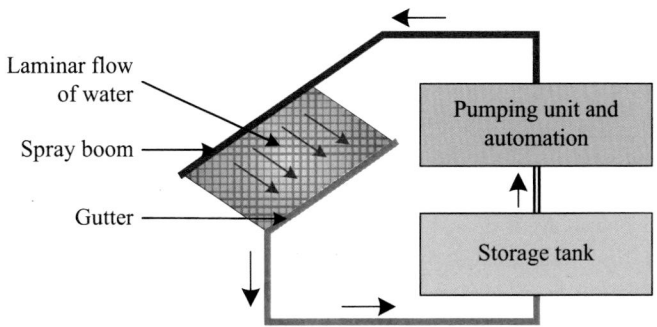

Fig. 5.50 Schematic diagram of the cooling system of a solar PV array

5.7   Conclusion

This chapter presents the components used in a PV system besides the solar PV array itself. The chapter begins with solar modules and then describes batteries, charge controllers, inverters, and other balance of system components such as wiring, switches, grounding and lightning protection systems, module mounting and tracking mechanisms, and cooling systems. For each component, the general description follows the detailed sizing procedure with a sizing example. The codes and standards by the National Electric Code 2020 are elaborately explained wherever necessary in this chapter. By the end of this chapter, the reader will have a solid knowledge about each individual component used in a solar PV system and will know how to choose the right type of components according to the application. In the next chapter, we will explore how to practically design PV systems utilizing the knowledge acquired from this chapter.

5.8   Exercise

1. Describe the different types of solar modules.
2. Discuss the comparison between series and shunt charge controllers.
3. Discuss the comparison between PWM and MPPT controllers.
4. What are the differences between a single-axis and dual-axis solar tracker?
5. How many types of inverters can be used in solar PV systems?
6. What are the different methods of mounting solar PV modules?
7. Describe the different tracking methods used in solar PV systems.
8. How are solar PV modules cooled?
9. What is the difference between EGC and GEC in grounding systems?
10. Define the following:

    - Step and touch potential
    - Net metering
    - Grounding
    - Balance of system
    - Charge controller
    - Float voltage
    - Days of autonomy
    - Depth of discharge
    - State of charge

11. Suppose a speaker inside a ship requires 400 Wh energy at 24 V. The ship is set to travel from Vancouver to San Francisco during cloudy weather. The cruise is expected to last for 4 days. A battery bank consisting of PG-12 V 60 Ah VRLA

Table 5.18 3-phase inverter specification sheet for Exercise 12

Nominal AC power	125 kW
Maximum apparent power	125 kVA
Nominal AC voltage	480 V
Maximum output current	151 A
Frequency	60 Hz

batteries with 50% DoD is considered to be installed to run the speaker during this period. The ambient temperature will be approximately 40 °F. Find out the number of batteries required to operate the speaker successfully. [Answer: 6 batteries; 2 modules in series, 3 strings in parallel]

12. Suppose a solar PV array consists of 16 parallel-connected strings, and each of those strings contains 46 series-connected modules. The modules have a V_{oc} of 22.5 V and I_{sc} of 8.9 A (Consider ambient temperature 60 °C). The system is using an SMA SUNNY HIGHPOWER PEAK3 125-US inverter (Specification sheet snippet provided in Table 5.18), which produces an output of 151 A. Determine:

- How the DC disconnect of this system should be rated?
- Mention the method of deciding on the AC disconnect current rating and calculate it.

[Answer: Minimum size of DC disconnect 277.68 A; AC disconnect 188.75 A]

13. Three parallel loads require 306 W, 220 W, and 156 W, respectively. A 24 V battery with 200 Ah is supporting the system. What would be the required power rating in VA of the inverter having a power factor of 0.7? How long can the battery provide backup power? Consider an ideal inverter with 100% efficiency in this case. [Answer: 974.28 VA; 7 h]

14. The performance of PV modules is the best in sunny days but cooler temperatures. On the other hand, the performance of batteries is better in warmer temperatures. So, in case of a solar plus storage system, how is the temperature impact balanced for these two components? State your opinion. Is the performance of PV systems better in the sunny belt (28 °C) or in the snow belt (14 °C) of the USA? Share your thoughts on this.

References

1. T.S. Dinh, M. Mussetta, G. Nicoleta Sava, M.Q. Duong, K.H. Le, Effects of bypass diode configurations on solar photovoltaic modules suffering from shading phenomenon, in *The 10th International Symposium on Advanced Topics in Electrical Engineering*
2. Fraunhofer Institute for Solar Energy Systems, Photovoltaics Report (2020). https://www.ise.fraunhofer.de/content/dam/ise/de/documents/publications/studies/Photovoltaics-Report.pdf
3. PV insights (2021). http://pvinsights.com/
4. R. Brakels, Half cut solar panels: higher efficiency & better shade tolerance (2018). https://www.solarquotes.com.au/blog/half-cut-solar-cells-panels/

5. X. Sun, M. Ryyan Khan, C. Deline, M. Ashraful Alam, Optimization and performance of bifacial solar modules: a global perspective. Appl. Energy **212**, 1601–1610 (2018)
6. S. Zhou, Y.-S. Jiang, P. Chen, G.-P. Huang, H. Zhuang, J.N. Li, Study of the twelve busbar technology and the stress-induced degradation within the solar modules. Adv. Eng. Res. **185**, 197–202 (2019)
7. REC, Guide to best practice—managing mismatches when replacing panels and using panels with different power ratings in a string (2014). https://www.irishellas.com/files/Managing_mismatches_when_replacing_panels_eng.pdf
8. Slobodan Petrovic, *Battery Technology Crash Course: A Concise Introduction* (Springer Nature, 2020)
9. E. Hossain, D. Murtaugh, J. Mody, H.M.R. Faruque, M.S.H. Sunny, N. Mohammad, A comprehensive review on second-life batteries: current state, manufacturing considerations, applications, impacts, barriers & potential solutions, business strategies, and policies. IEEE Access **7**, 73215–73252 (2019)
10. E. Hossain, H.M.R. Faruque, M.Sunny, S. Haque, N. Mohammad, N. Nawar, A comprehensive review on energy storage systems: types, comparison, current scenario, applications, barriers, and potential solutions, policies, and future prospects. Energies **13**(14), 3651 (2020)
11. D.G. Enos, Lead-acid batteries and advanced lead-carbon batteries. Technical report, Sandia National Lab.(SNL-NM), Albuquerque, NM (United States), 2014
12. B.B. McKeon, J. Furukawa, S. Fenstermacher, Advanced lead–acid batteries and the development of grid-scale energy storage systems. Proc. IEEE **102**(6), 951–963 (2014)
13. S. Anuphappharadorn, S. Sukchai, C. Sirisamphanwong, N. Ketjoy, Comparison the economic analysis of the battery between lithium-ion and lead-acid in PV stand-alone application. Energy Proc. **56**, 352–358 (2014)
14. H. Qiao, Q. Wei, 10—functional nanofibers in lithium-ion batteries, in *Functional Nanofibers and Their Applications*, ed. by Q. Wei. Woodhead Publishing Series in Textiles (Woodhead Publishing, 2012), pp. 197–208
15. J. Li, C. Ma, M. Chi, C. Liang, N.J. Dudney, Solid electrolyte: the key for high-voltage lithium batteries. Adv. Energy Mater. **5**(4), 1401408 (2015)
16. G. Fabbri, M. Paschero, A.J.M. Cardoso, C. Boccaletti, F.M. Frattale Mascioli, A genetic algorithm based battery model for stand alone radio base stations powering, in *2011 IEEE 33rd International Telecommunications Energy Conference (INTELEC)* (IEEE, Piscataway, 2011), pp. 1–8
17. V. Aravindan, J. Gnanaraj, Y.-S. Lee, S. Madhavi, LiMnPo4—a next generation cathode material for lithium-ion batteries. J. Mater. Chem. A **1**(11), 3518–3539 (2013)
18. BU-107: Comparison table of secondary batteries (2021). https://batteryuniversity.com/article/bu-107-comparison-table-of-secondary-batteries
19. EESemi, Electrode reduction and oxidation potential (2021). https://www.eesemi.com/ox_potential.htm
20. W. Waag, D. Sauer, *Secondary Batteries—Lead-Acid Systems | State-of-Charge/Health* (2009), pp. 793–804
21. P. Zhang, J. Liang, F. Zhang, An overview of different approaches for battery lifetime prediction, in *IOP Conference Series: Materials Science and Engineering*, vol. 199 (IOP Publishing, 2017), p. 012134
22. D. Spiers, Batteries in PV systems, in *Practical Handbook of Photovoltaics* (Elsevier, Amsterdam, 2012), pp. 721–776
23. Characteristics of lead acid batteries (2021). https://www.pveducation.org/pvcdrom/lead-acid-batteries/characteristics-of-lead-acid-batteries
24. IEEE recommended practice for sizing lead-acid batteries for stationary applications. *IEEE Std 485-2020 (Revision of IEEE Std 485-2010)* (2020), pp. 1–69
25. H.J. Khasawneh, A. Mondal, M.S. Illindala, B.L. Schenkman, D.R. Borneo, Evaluation and sizing of energy storage systems for microgrids, in *2015 IEEE/IAS 51st Industrial & Commercial Power Systems Technical Conference (I&CPS)* (IEEE, Piscataway, 2015), pp. 1–8
26. C. Carriero, Battery stack monitor maximizes performance of Li-ion batteries in hybrid and electric vehicles

27. H. Hemi, J. Ghouili, A. Cheriti, Dynamic modeling and simulation of temperature and current effects on an electric vehicles lithium ion battery, in *2015 IEEE 28th Canadian Conference on Electrical and Computer Engineering (CCECE)* (IEEE, 2015), pp. 970–975

28. E. Chiodo, D. Lauria, N. Andrenacci, G. Pede, Probabilistic battery design based upon accelerated life tests. Intell. Ind. Syst. **2**(3), 243–252 (2016)

29. S. Caux, of deliverable: advanced source and material profiling report v2. Energy **56**, 64–71 (2013)

30. S. Jena, Design and analysis: 25 kw standalone PV system (2019)

31. Electrical Academia, Battery charger types | trickle & float charger working (2021). https://electricalacademia.com/batteries/battery-charger-types-trickle-float-charger-working/

32. CZH-LABS LVD low voltage disconnect module. https://www.amazon.com/CZH-LABS-LVD-Voltage-Disconnect-Module/dp/B07WDLFFFB?th=1

33. C.A. Osaretin, F.O. Edeko, Design and implementation of a solar charge controller with variable output. Electr. Electron. Eng. **12**(2), 40–50 (2015)

34. Solar Pro, Understanding and optimizing battery temperature compensation (2021). http://origin-faq.pro-face.com/resources/sites/PROFACE/content/live/FAQS/229000/FA229659/en_US/SP3.1%20Goodnight%20Battery%20Temp%20Comp.pdf

35. K. Jlassi, Battery temperature compensation (2018). https://www.sunwize.com/battery-temperature-compensation/#:~:text=V%2F%C2%B0C.-,The%20temperature%20compensation%20value%20is%20from%2025%C2%B0C%2C%20so,C%20would%20be%20~14.4V

36. Clean Energy Project, PWM vs MPPT solar charge controller: which one to choose? (2021). https://cleanenergyprojectnv.org/pwm-vs-mppt/

37. RENOGY, Solar charge controller sizing and how to choose one (2021). https://www.renogy.com/blog/solar-charge-controller-sizing-and-how-to-choose-one-/

38. HQ SOLAR POWER, Charge controller types (2021). https://hqsolarpower.com/learn-charge-controller-types/

39. SMA, Sunny Central (2021). https://files.sma.de/downloads/SC2200-3000-EV-DS-en-59.pdf

40. SOLAR STIK, Calculating inverter size (2021). https://www.solarstik.com/stikopedia/calculate-inverter-size/

41. Learning Electrical Engineering, How to calculate inverter power rating and inverter battery backup time (2021). https://www.electricalengineeringtoolbox.com/2017/07/how-to-calculate-inverter-power-rating.html

42. American wire gauge (AWG) cable/conductor sizes (2021). https://diyaudioprojects.com/Technical/American-Wire-Gauge/

43. Anixter, Copper vs. aluminum conductors (2020). https://www.anixter.com/en_au/resources/literature/wire-wisdom/copper-vs-aluminum-conductors.html

44. Monroe Engineering, Copper vs aluminum wiring: Which is best? (2020). https://monroeengineering.com/blog/copper-vs-aluminum-wiring-which-is-best/

45. National Fire Protection Association (NFPA), *NFPA 70—National Electric Code* (2020)

46. Solar Energy International, *Photovoltaics: Design and Installation Manual* (2004)

47. B. Brooks, S. White, *Photovoltaic Systems and the National Electric Code* (Routledge, London, 2018)

48. Solar Energy International, *Solar Electric Handbook* (Pearson, 2013)

49. ABB, Disconnect switches—applications in photovoltaic systems (2021). https://search.abb.com/library/Download.aspx?DocumentID=1SXU301197B0201&LanguageCode=en&DocumentPartId=&Action=Launch

50. Schneider Electric, Correction factor table for powerpact b-frame circuit breakers (2021). https://www.productinfo.schneider-electric.com/powerpactb/59df870a46e0fb000184e952/PowerPact

51. ZJBENY, DC disconnect switch 1000v 400a bh-400 IEC (2021). https://www.beny.com/product-item/dc-disconnect-switch-1000v-400a-bh-400/

52. SMA, Sunny boy 3.0/3.6/4.0/5.0/6.0 (2021). https://files.sma.de/downloads/SB30-60-DS-en-42.pdf

53. Siemens WF2030 30 amp fusible AC disconnect (2021). https://www.amazon.com/Siemens-WF2030-Amp-Fusible-Disconnect/dp/B0052MB01O
54. The Solar Planner, Photovoltaic tutorial: Step-by-step guide to going solar (2019). https://www.thesolarplanner.com/steps_page8.html
55. CED GREENTECH, Sizing the DC disconnect for solar PV systems (2021). https://www.cedgreentech.com/article/sizing-dc-disconnect-solar-pv-systems
56. Energy Sage, What is net metering? (2021). https://www.energysage.com/solar/solar-101/net-metering/
57. A. Awasthi, A.K. Shukla, S.R. Murali Manohar, C. Dondariya, K.N. Shukla, D. Porwal, G. Richhariya, Review on sun tracking technology in solar PV system. Energy Rep. **6**, 392–405 (2020)
58. Solar Feeds, Types of solar trackers and their advantages & disadvantages (2021). https://www.solarfeeds.com/mag/solar-trackers-types-and-its-advantages-and-disadvantages/
59. K.A. Moharram, M.S. Abd-Elhady, H.A. Kandil, H. El-Sherif, Enhancing the performance of photovoltaic panels by water cooling. Ain Shams Eng. J. **4**(4), 869–877 (2013)
60. E. Bellini, Cooling down PV panels with water (2020). https://www.pv-magazine.com/2020/03/31/cooling-down-pv-panels-with-water/

Standalone, Hybrid, and Distributed PV Systems

6

6.1 Introduction

Based on grid connectivity, solar PV systems are of three types: grid-tied PV system, off-grid or standalone PV system, and hybrid PV system. In this chapter, the design processes of standalone and hybrid PV systems are described. Grid-tied PV systems will be explained in Chap. 7. Again, based on the size and application of the system, solar PV systems can be either utility-scale solar PV systems or distributed solar PV systems. In this chapter, the design processes of two distributed solar PV systems— solar water pumping and street lighting systems—are also explained.

6.2 Off-Grid PV Systems

Standalone or off-grid systems are very popular, especially in developing countries, due to the absence of a reliable utility grid or prolonged electricity outages. These systems are also widely used in remote areas. They have multiple applications, including, but not limited to, residential or service applications (health centers, school, condominium), street lighting systems, bus station equipment, water pumping, telecommunication sites, refugee camps, traffic control devices (traffic lights). Solar technology provides an excellent alternative solution for these cases due to the high costs of expanding the utility grid. This technology is more effective than building new substations and transmission lines. Off-grid PV systems are more feasible since diesel or other oil products to run small generators can be financially and environmentally costly.

Although the PV modules do not produce energy at night, off-grid PV systems can ensure reliable energy throughout the year because of the energy storage systems (usually batteries) that are sized based on the importance of the loads. For instance, telecommunication sites have very critical devices to receive and send signals in remote areas. These devices should be supplied with reliable power, independent of any atmospheric conditions, such as snow, clouds, or storms. Therefore, this type of system's PV and energy storage is oversized and more frequently replaced than any other application to ensure a continuous and reliable supply.

On the contrary, water pumping systems do not usually rely on oversizing the energy storage system but on water storage by oversizing the PV and pump power rating. This difference in sizing prevents energy conversion losses, which vary between 10 and 20% based on the battery technology.

The number of autonomy days is when the energy storage system can supply the entire load without any assistance from the PV arrays. This number of days varies from 0 to 15 days. The more the days of autonomy, the more critical is the load and the more expensive the system is. The daily load demand has to be multiplied by the days of autonomy to get the required battery capacity. However, the days of autonomy are determined based on the following two main factors:

1. **Funding parties' budget:** Energy storage components are the most expensive in any off-grid application. Their cost usually accounts for 40–50% of the initial cost, and they need replacement once their lifespan ends. These expenses might exceed the client or the funding parties' budget, which reduces autonomy days or shortens the energy supply period.
2. **Geographical or mechanical limitations:** In some cases, the surface on which the modules should be mounted might not face the right direction (south on the north hemisphere or north on the south hemisphere). It might also be exposed to soft shadings (such as tree branches) or hard shadings (such as HVAC fans or crowns). In parallel, existing street lighting poles might not be designed and structured to handle 20–40 kg weighted solar modules. Therefore, the designer should consider the mechanical strength of the poles or the roof in residential applications.

The non-redundant components of any battery-based standalone system contain solar modules, batteries, charge controller, cables, protection devices, DC and/or AC loads, and inverters (in case of AC loads' availability). Therefore, the system sizing process can be broken down into several sub-steps through which each component is sized to have an accurate balance between the load demand and energy yield. The losses and limitations, such as voltage drop, battery inefficiencies, and depth of discharge, and most importantly, meeting the safety standards, are not considered in this process for ease of calculations.

6.2.1 System Sizing Process

For an off-grid solar PV system, modules, battery, charge controller, and inverter are the most important components. System sizing involves the determination of the sizes of these components based on the required energy needs. There are seven steps in the system sizing process. They are discussed step-by-step in the following sections.

6.2.1.1 Load Analysis and Evaluation

The sizing process is started with determining the load type and profile through collecting the data of the power ratings of each electric device and its operating time to obtain the daily energy consumption of the application. The following ratings are collected:

1. **Power rating:** It is the highest nominal power that can be fed to the input of the electric device, measured in Watt. The following equation is used to find the nominal power:

$$P = V \times I. \tag{6.1}$$

2. **Energy consumption:** It is measured in Watt-Hour and found by multiplying the power rating and the operating time. The following equation shows that the energy E is the product of the power P and the time t:

$$E = P \times t. \tag{6.2}$$

3. **Load factors:** The load factor (F_{load}) is the ratio of the total energy consumption for a specific number of hours to the maximum power rating and the number of operating hours as shown in Eq. 6.3. The value is between 0 and 1. For instance, a refrigerators' compressor does not work 100% all the time, and when it operates, it always works on maximum power rating. Typically, it operates for only 40% of the day. In other words, the average load factor of the refrigerator is 40%.

$$F_{load} = \frac{Energy\ consumed}{Maximum\ power \times Operating\ hours}. \tag{6.3}$$

4. **Battery energy:** The total energy needed from the batteries to supply the DC and AC loads can be calculated using the following equation:

$$E_{batteries} = E_{DC\ load} + \frac{E_{AC\ load}}{\eta_{inverter}}. \tag{6.4}$$

Let us now have a look at an example that describes the analysis of the daily loads of a house.

Example 6.1 (Load Analysis for DC Loads) Determine the daily energy demand of the following DC loads:

Table 6.1 Table for calculating the daily energy consumption of DC loads in Example 6.1

Device	Quantity	Power rating (W)	Operating time (h)	Load factor	Energy consumption (Wh)
Fluorescent lamps	10	40	12	1	4800
Refrigerator	1	500	24	0.4	4800
Laptop chargers	5	60	5	1	1500

1. Ten fluorescent lamps operate for 12 h/day. Each of them has a 40 W power rating.
2. One refrigerator has a power rating of 500 W. The load factor is 40%.
3. Five laptop chargers operate 5 h per day with a power rating of 60 W for each.

Solution The daily energy demand of the given DC loads is shown in Table 6.1.

The daily energy consumption of these loads is 11,110 Wh. It is essential to remember that most loads are run using AC power since the utility grid is an AC source. Therefore, an AC inverter should convert the DC energy generated by the PV modules or stored by the batteries to AC power when the application contains AC loads. The DC/AC energy conversion causes some inefficiencies, which vary between 5 and 10%, and this should be considered in the calculation process.

6.2.1.2 Determination of System Voltage

As discussed in Chap. 5, lead-acid batteries have a nominal voltage of 12 V, comprising six 2 V cells connected in series. The batteries are connected in series to produce a higher voltage. All batteries are classified by their voltage (V) and ampacity (A). The nominal system voltages in off-grid systems are 12, 24, and 48 V. A 12 V battery is usually used with systems that do not contain more than 750 W devices as a cumulative operating power rating. At the same time, a 24 V battery can handle up to 1500 W, and a 48 V battery is designed for systems up to 3000 W. In the series connection of batteries, the amount of current stays the same.

It is essential to mention that these numbers are also dictated by the power rating of the DC devices. For instance, if the DC electric devices of the system have a power rating of 12 V, a designer should design the system based on the nominal operating voltage rating of the devices. Otherwise, buck or boost converters (buck converters reduce the voltage, whereas boost converters increase the voltage) will be required to transfer the battery voltages to the operating voltage, thereby increasing the losses and costs.

Using a higher voltage reduces the current flowing in the cables significantly. This reduces the cross-section of the cable along with the Ohmic losses. So, the higher the system voltage, the lower is the system current; hence, the lower is the system loss. For instance, if a 48 V system is used instead of a 12 V system for the same load, then the current will decrease by $12 \div 48 = 25\%$. Again, if a 24 V system is used, the current will decrease to $24 \div 48 = 50\%$.

6.2.1.3 Battery Bank System Sizing

After determining the system voltage, the designer can convert the energy demand measured in Wh into battery ampacity measured in Ah. In this step, the losses in the battery discharge process should be taken into consideration along with the maximum allowable depth of discharge (DoD) that the battery cannot exceed to maintain the battery's good operating conditions. Lead-acid batteries do not exceed 80% DoD, while lithium-ion can be fully discharged. It is essential to remember that the higher the DoD, the lesser the lifespan of the batteries, but lesser the initial cost, as implied by the following equation:

$$Battery\ Bank\ Capacity = \frac{E_{critical\ month} \times Days\ of\ Autonomy}{DoD \times Battery\ system\ voltage}. \quad (6.5)$$

Here, $E_{critical\ month}$ is the total daily energy consumption during the critical month. The critical month refers to the month in which the solar irradiation is the least in the year. This information can be obtained from the weather and climate data from the meteorological office of any location. This month is considered because it is when the necessity of the battery bank is the highest. The consideration of the maximum required capacity of the battery bank helps to optimize the calculation and helps in the appropriate sizing of the battery.

Figure 6.1 illustrates how the battery bank capacity varies with temperature at different discharge rates. A fast discharge will cause the battery capacity to be lower than that of a slower discharge. In outdoor applications, the designer should be aware of the temperature effect on the battery, which is incorporated in the following equation:

$$Battery\ bank\ capacity_{with\ temperature\ effect} = \frac{Battery\ bank\ capacity}{TCC}, \quad (6.6)$$

where

TCC is the temperature capacity coefficient.

Example 6.2 (Battery Bank Sizing) In addition to the previous loads of 11,110 Wh in Example 6.1, calculate the energy consumption of the AC loads in Table 6.2 supplied by a sine-wave inverter with an efficiency of 92%. Also, calculate the total energy needed from the batteries to supply both AC and DC loads.

Solution The daily energy demand of the given AC loads is shown in Table 6.2. The total energy that is needed from the batteries to supply the DC and AC loads can be determined using the following equation:

$$E_{batteries} = E_{DC\ load} + \frac{E_{AC\ load}}{\eta_{inverter}} = 11,110 + \frac{2880 + 7200}{0.92}$$

$$= 22,056.5\ Wh.$$

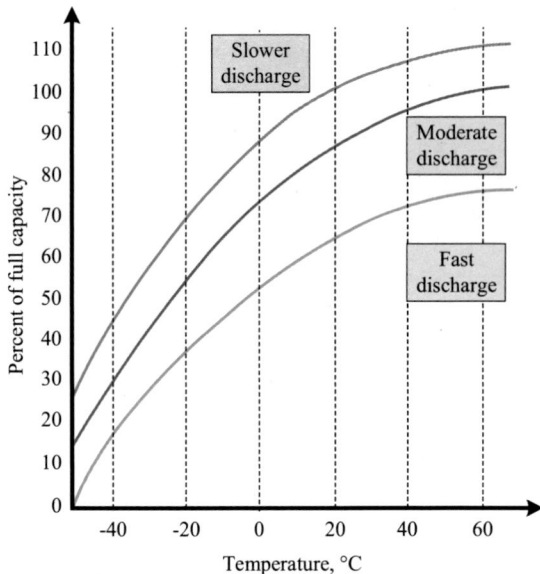

Fig. 6.1 Change of capacity of battery bank with the rise of temperature [1]

Table 6.2 Table for calculating the daily energy consumption of AC loads in Example 6.2

Device	Quantity	Power rating (W)	Operating time (h)	Load Factor	Energy consumption (Wh)
Washing machine	1	2400	2	0.6	2880
LED screen	2	300	12	1	7200

Due to the atmospheric and temperature changes that might occur during the year, the loads and operating time might change. Therefore, the system should be designed to meet the worst-case scenario's load requirements that have the highest load demand with the lowest expected irradiance. Usually, solar irradiance is the lowest during the winter. Therefore, solar energy is not sufficient to run the required loads, and the dependence on the battery bank is the highest during the winter. So, most applications are designed to meet the winter load case unless the load is tremendously boosted up during the other months of the year.

6.2.1.4 Evaluation of Solar Resources

As mentioned earlier, PV modules should be tilted toward the sun by an optimum tilt angle that maximizes the solar irradiance exposure on the non-concentrating solar collectors in the critical month or season. Most resources provide the values of global horizontal irradiance (GHI). However, a designer should plan the system considering the global tilted irradiance (GTI). The energy yield of a solar module depends on the tilt angle and GTI as well, for which considering this calculation while planning is essential to have an accurate estimation. The MATLAB code given in Sect. 4.16 can calculate the GTI from GHI values at any place in the world.

6.2.1.5 Determination of PV Array Size

The PV array size varies depending on the efficiency of energy and the energy usage of the place where it is installed. Solar modules of the system should be sized to provide enough energy to charge the batteries and cover the energy losses in the charging process (known as battery efficiency). So, the next step in the sizing process is determining the PV array size, which includes finding the power rating and current of the solar modules by finalizing the voltage of the battery system. The power rating of the solar modules can be calculated using the following equation:

$$P1_{PV\ array} = \frac{E_{critical\ month}}{\eta_{batteries} \times GTI_{critical\ month}}, \tag{6.7}$$

where

$E_{critical\ month} =$ daily energy usage in critical month,
$\eta_{batteries} =$ efficiency of the battery,
$GTI_{critical\ month} =$ value of GTI in that month.

As discussed in Sect. 2.15, a solar module has to go through some specific tests before entering the market, which includes testing the temperature coefficient and ability to tackle outdoor exposure. So, it is comprehensible that the environment has a major impact on the performance of the solar system. For this reason, temperature and soiling losses should also be considered in the calculation process to avoid an energy supply shortage. Therefore, the final power rating of the solar modules considering the effect of temperature and soiling can be calculated using the following equation:

$$P_{PV\ array} = \frac{P1_{PV\ array}}{\eta_{thermal} \times \eta_{dirt}}, \tag{6.8}$$

where

$\eta_{thermal}$ is the efficiency considering the thermal losses, which depends on the highest ambient temperature, and
η_{dirt} is the efficiency considering the dirt losses.

6.2.1.6 Charge Controller Sizing

The selection process of charge controllers should be based on two main criteria. First, the nominal voltage output of the charge controller should be compatible with the battery bank system voltage. Second, the controller should handle 125% of the cumulative short circuit of the PV array, so that the device does not have to carry more than $1 \div 1.25 = 80\%$ of its capacity. New inverters have one or multiple charge controllers embedded in the inverter casing with multiple inputs, such as the 5 kVA 4 kW Hybrid Pure Sine Wave Inverter with 60 A 48 V MPPT Controller by Vikye [2]. However, usually, these inverters are not equipped with DC loads. In such cases, the user should use a rectifier for AC–DC conversion of some output power to supply the DC loads.

6.2.1.7 Inverter Sizing
For selecting a suitable inverter for the solar PV system, the following conditions should be met:

- The input nominal DC voltage should be equal to the battery bank voltage.
- The frequency and nominal AC output voltage should be equal to the frequency and nominal AC output voltage of the AC loads.
- The inverter's power rating should be higher than the load's maximum cumulative power that is expected to work simultaneously.
- Load characteristics should be checked in order to select the output waveform.
- In the case of using induction motors and/or sensitive machinery, a pure sine wave inverter should be used.
- In the case of using motors, the inverter should handle the surge or inrush current of the motor that varies from 2 to 5 times of the nominal operating current for a very short time. It is worth noting that most of the modern inverters are designed to handle the expected inrush currents of the induction motors.
- When using the variable frequency drive (VFD) technology to control the motor speed, the inrush current can be neglected.

6.2.2 Case Study: Indonesia

A house in Batam, Indonesia, has a 35° tilted roof on which solar PV modules could be installed to satisfy the needs of its 2.2 kW appliances that have the daily average load profile provided in Table 6.3. Using the previously mentioned steps, we are going to design a standalone PV system, which should supply energy to the loads for 2 days without any solar support, i.e., have 2 days of autonomy. The details of the batteries and the modules are in Tables 6.4 and 6.5. We assume that the inverter efficiency is 95%.

For the mentioned case study, the steps of designing the PV system are described as follows:

1. **Load Analysis and Evaluation:** From Table 6.3, we get the total average daily energy consumption in the house as 15,926 Wh. Therefore, the energy that the battery bank should supply to the load can be found from the following equation:

$$E_{batteries} = E_{DC\,load} + \frac{E_{AC\,load}}{\eta_{inverter}} = 0 + \frac{15,926}{0.95} = 16,765\ Wh.$$

Table 6.3 Daily load profile
of the house in Indonesia

Hour of day	Energy, Wh
0–1	1590
1–2	650
2–3	180
3–4	180
4–5	180
5–6	430
6–7	475
7–8	360
8–9	110
9–10	250
10–11	360
11–12	460
12–13	302
13–14	292
14–15	152
15–16	360
16–17	250
17–18	600
18–19	1085
19–20	1730
20–21	1490
21–22	1450
22–23	1400
23–24	1590
Total	15,926

Table 6.4 Specification of
PS-P60-(275-290 W)
Polycrystalline Module by
Philadelphia Solar [3]

Maximum power rating (mpp)	275 W
Open-circuit voltage (V_{oc})	38.31 V
Short-circuit Current (I_{sc})	9.04 A
Maximum Voltage (V_{mpp})	32.13 V
Maximum Current (I_{mpp})	8.56 A
Module efficiency	16.76%
Power temperature coefficient	−0.4%/°C

Table 6.5 Battery
specification

Battery technology	Lead-acid battery
Nominal voltage	12 V
Ampacity	100 Ah
Maximum DoD	80%
Charge–discharge efficiency	90%

2. **Determination of System Voltage:** Since the total power rating of the appliances used in the house is 2.2 kW, the appropriate system voltage is 48 V.

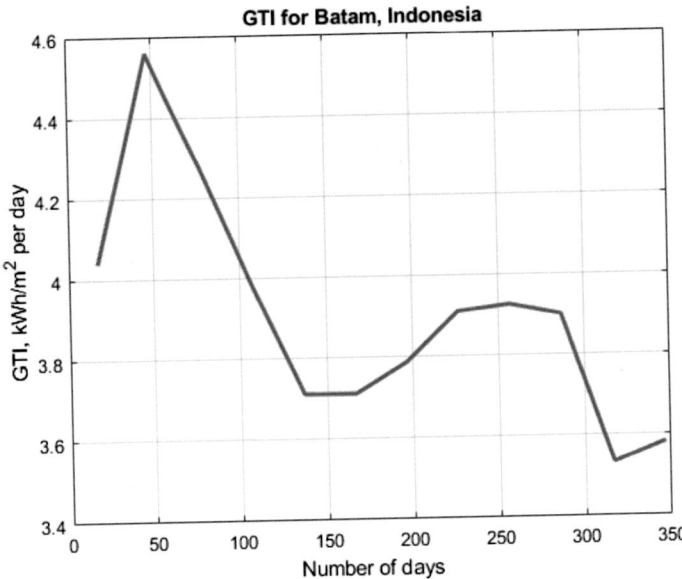

Fig. 6.2 GTI for the house in Batam, Indonesia

3. **Battery Bank System Sizing:** Since the system voltage is 48 V, it is accomplished by four 12 V batteries connected in series, where $12 \times 4 = 48$ V. The battery bank capacity at 48 V can be found by the following equation:

$$Battery\ Bank\ Capacity = \frac{E_{batteries} \times Days\ of\ autonomy}{DoD \times System\ voltage}$$

$$Battery\ Bank\ Capacity = \frac{16,765 \times 2}{0.8 \times 48} = 873\ Ah.$$

Since the battery bank will be installed indoors, the temperature–capacity coefficient might be neglected due to the low discharge capacity, and the operating temperature is near 25 °C.

4. **Evaluation of Solar Resources:** Global horizontal irradiance (GHI) can be found from several sources. The GHI values are used as input values to the MATLAB code in Sect. 4.16 to determine the GTI at the tilted surface on which the solar modules are installed. Figure 6.2 illustrates the monthly GTI in this location. The lowest value of GTI, i.e., 3.55 is used as the worst-case scenario at which the system is supposed to be designed.

5. **Determination of PV Array Size:** Since the critical month and average energy consumption are determined, we can calculate the PV array sizing that will charge the batteries using the following formula:

$$P1_{PV\ array} = \frac{E_{critical\ month}}{\eta_{batteries} \times GTI_{critical\ month}} = \frac{16,765}{0.9 \times 3.55} = 5247\ W.$$

Temperature and soiling losses should also be considered in the calculation process in order to avoid any energy supply shortage. Here, the cell temperature is $60\,°C$. The power coefficient is $-0.4\%/°C$. For $60 - 25 = 35\,°C$ temperature rise, the thermal power loss is $35 \times (-0.4) = -14\%$. Thus, the thermal efficiency $\eta_{thermal} = 100 - 14 = 86\%$. Again, the dirt loss is 2%, so $\eta_{dirt} = 100 - 2 = 98\%$.

$$P_{PV\ array} = \frac{P1_{PV\ array}}{\eta_{thermal} \times \eta_{dirt}} = \frac{5247}{0.86 \times 0.98} = 6226\ W.$$

6. **Charge Controller Sizing:** The number of modules can be obtained by dividing the PV system power rating by the nominal power rating of each module:

$$\frac{6226\ W}{275\ W} = 22.64 \approx 24\ modules.$$

This number is rounded up to 24 since the number should be even due to the 48 V system preferred for the PV array. The PV module configuration is $12 \times 2 = 24$ modules since the battery system voltage is 48 V. So, 2×32.13 (from Table 6.4) $= 64.26$ V is supplied by the modules by connecting 12 strings in parallel. Each string contains two modules connected in series. It is worth noting that this module configuration is selected based on a conventional charge controller that does not use MPPT technology. If the MPPT technology is used, the PV array voltage could be much higher than the voltage of choice; this would allow using a smaller cable size, thus reducing the costs. The selection of the charge controller is based on the short-circuit current of the array:

$$9.04\ A \times 12 = 108.48\ A.$$

Therefore, the charge controller ampacity should be at least $108.48 \times 1.25 = 135.6$ A.

7. **Inverter Sizing:** Since all the home appliances work on AC and their total power rating is 2.2 kW, the inverter should generate pure sine wave voltage signals with a power rating of no less than 2.2 kW. It is essential to keep in mind that the input nominal voltage is 48 V, and the output signal satisfies the Indonesian standards of electrical devices (i.e., 220 V, 50 Hz).

6.3   Hybrid PV Systems

A grid-tied PV system does not usually contain a battery backup since all the surplus power is sent to the utility grid. A hybrid PV system is a grid-tied PV system, but it contains a battery energy storage system (BESS) for storing surplus power. Often nicknamed "solar plus storage systems," such systems contain three sources of power for any load: the utility grid, the solar PV arrays, and the battery bank. In such a system, solar energy is prioritized as the energy source, followed by the utility grid and the BESS. If solar energy is unavailable, then the utility grid or the BESS is weighed in terms of costs. If using the grid power is cheaper, then energy is consumed from the grid, and if the grid power is expensive (for example, at peak hours), then the BESS is used as the energy source. The block diagram of a hybrid solar PV system is depicted in Fig. 6.3.

6.3.1   System Sizing

System sizing refers to the determination of the size of the PV system required to meet the load requirements of the facility. For designing a PV system for a residential building, the total daily loads of the system should be taken into consideration. Let us consider that a certain application requires 5 kWh of energy per day, where the maximum power rating does not exceed 2 kW, and plan to design a solar plus storage system for the application.

6.3.2   Site Assessment

The site assessment involves two prime considerations: information of the site of installation of the PV array and the solar resource allocation at the site to obtain

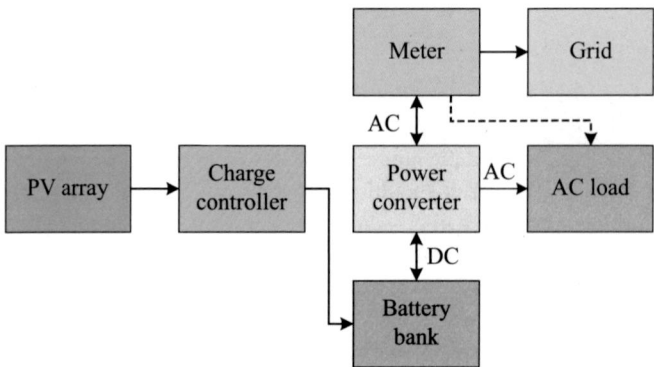

Fig. 6.3 Schematic of a hybrid solar PV system. The charge controller is not required if the system uses a hybrid inverter with a built-in charge controlling unit

optimum performance. Say the application mentioned in this design study is located in Oregon, USA. In that case, we would require the sun path diagram for Oregon, which is provided in Fig. 4.12. This diagram provides a peek into the solar window available at the said location from June through December. The latitude of Oregon is 43.8041° North. So, the desired tilt angle of the PV arrays would be around 44°, and the modules should be facing the south. It is customary to consider 30% losses in the system, and so the module should be sized large enough to compensate for the losses. So, the PV module size is multiplied by 1.3 to get the actual required size. If the location has approximately 4 peak sun hours (PSH), then the required PV module size is:

$$(5000\ Wh \div 4\ h) \times 1.3 = 1625\ W.$$

6.3.3 PV Array Sizing

The size of the PV array depends on the system size. For the mentioned application, the solar module can be a monocrystalline, 60-cell, 380 W module. The monocrystalline module is chosen for its higher efficiency compared with polycrystalline modules. In addition, a 60-cell module is sufficient for residential applications. The specifications are given in Table 6.6.

$$No.\ of\ PV\ modules = \frac{1625\ W}{380\ W} = 4.27 \approx 5.$$

So, 5 PV modules have to be connected in series. The output open-circuit voltage of the PV array $= 5 \times 47.95\ V \approx 240\ V.$

Table 6.6 Q.PEAK DUO L-G5.2 380–395 module specification for the hybrid system [4]

Parameter	Value
Open-circuit voltage, V_{oc} (V)	47.95
Short-circuit current, I_{sc} (A)	10.05
Maximum power voltage, V_{mpp} (V)	39.71
Maximum power current, I_{mpp} (A)	9.57
Maximum power, P_{max} (W)	380
Module efficiency, %	18.9
Voltage temperature coefficient, (%/°C)	−0.28
Current temperature coefficient, (%/°C)	+0.04
Power temperature coefficient, (%/°C)	−0.27
Dimensions (mm)	2015 × 1000 × 35

6.3.4 Battery Bank Sizing

The size and type of the battery bank depend on the days of autonomy and the load size. The voltage rating of the solar module and the battery should be compatible because they are connected directly with a disconnect between them. Ideally, the voltage rating of the module is higher than that of the battery, so that the current can flow from the higher potential to the lower potential. We consider 2 days of autonomy for the design considerations. This implies that the battery bank would be expected to provide backup to the loads for at least 2 whole days without any other source of power charging it or sharing the loads. So the required capacity would be $5000 \times 2 = 10,000$ Wh. Most battery banks are rated at 12, 24, or 48 V. Suppose that we are using 12 V batteries rated at 250 Ah, 85% efficient, with 50% DoD, such as a Mighty Max ML8D Sealed lead-acid rechargeable battery [5].

$$Battery\ bank\ capacity = \frac{Daily\ consumption}{DoD \times Efficiency \times System\ voltage}$$

$$= \frac{10,000\ Wh}{0.5 \times 0.85 \times 48\ V} = 490.2\ Ah.$$

Here, the system voltage is taken as 48 V because the output voltage of the PV module is about 48 V, and this voltage is necessary for the 3 kW system. Each battery is rated at 250 Ah; so, we would require $490\ Ah \div 250\ Ah = 1.96 \approx 2$ batteries connected in parallel to supply the required Ah capacity. Again, each battery is rated at 12 V; so, we would require 4 batteries in series to supply the 48 V of system voltage. Therefore, the total number of batteries is $4 \times 2 = 8$, with two rows of four batteries in series.

6.3.5 Inverter Sizing

The size of the inverter depends on the PV power output and the desired AC power deliverable by the overall system. For a hybrid system, a hybrid inverter or a power converter would be necessary. A hybrid inverter has multiple roles in such a system and performs both DC/AC and AC/DC power conversions. It can convert the DC PV power into AC power, which can be used by the connected loads. It also synchronizes the power for grid integration. Besides, the hybrid inverter can convert the AC power from the grid into DC power to feed the battery bank in case of solar power unavailability. A special advantage of a hybrid inverter is that it has a built-in charge controller, preventing the overcharging of the batteries and eliminating the need for a separate charge controller.

The total power rating of the five 380 W solar modules (from Sect. 6.3.3) is $5 \times 380 = 1900\ W$. The rated power of the inverter should be more than 1900 W for this system. While selecting the inverter, it is equally important to ensure that both

Table 6.7 Specification of the IGrid VE II 3.2 kW 48 V Hybrid Solar Inverter by EASUN Power for the hybrid PV system [6]

Parameter	Value
Maximum PV array power (W)	5500
Rated output power (W)	3200
MPPT range at operating voltage (V_{DC})	120–450
Grid-tied operation	
Grid output (AC)	
Nominal output voltage (V_{AC})	220/230/240
Output voltage range (V_{AC})	184–265
Nominal output current (A)	14.5/13.9/13.3
Maximum efficiency	93%
Off-grid hybrid operation	
Grid input	
Acceptable input voltage range (V_{AC})	120–280 or 170–280
Frequency range (Hz)	50/60 (auto sensing)
Battery mode output (AC)	
Nominal output voltage (V_{AC})	220/230/240
Output wave form	Pure sine wave
Battery and charger	
Nominal DC voltage (V_{DC})	48
Maximum AC charge current (A)	60
Maximum charge current (AC+PV) (A)	90
Emergency output power	
Maximum output power (W)	3200
Surge power (W)	6400
Automatic transfer time (ms)	<10

the voltage and current ratings of the inverter input side are compatible with the PV module.

A hybrid inverter with an output power of 3.2 kW will be enough for this system as it has a maximum PV input current of 30 A (maximum charge current 90 A—maximum AC charge current 60 A), which is greater than the output current of the PV array. For the chosen inverter, the output current is 14.5 A. Let us have a look at the 3.2 kW inverter whose specification is provided in Table 6.7. This size agrees with the PV array's voltage and current.

6.3.6 Wiring

6.3.6.1 PV Module to Inverter
The PV string has a short-circuit current of 10.05 A, and the cable ampacity should be rated at 156% of the short-circuit rating. So, the cable should be sized for at least $10.05 \times 1.56 = 15.678$ A. So, as per Table 310.16 of NEC 2020, a Copper THHW

cable of size AWG #14 would suffice, which can handle up to 20 A of current at 75°C.

But if the temperature reaches 46–50 °C, then the wire will only handle 20 × 0.75 = 15 A of current, according to Table 310.15(B)(1) of NEC 2020. So, the cable that can carry at least 15 A at 75% of its capacity is to be chosen.

In that case, AWG #14 at 75 °C will be the best choice, which can carry 20 A of current at normal temperature and 20 × 0.75 = 15 A at high temperatures, which suits the requirements.

The maximum voltage output from the 5 series modules is 39.71 × 5 = 198.55 V. So, according to the design criteria, the maximum voltage drop in the wire can be 2% of 198.55 V = 3.971 V.

Suppose that the distance between the PV module and the inverter is 30 ft. So, the two-way wiring will use up to 30 × 2 = 60 ft = 0.06 kft of wire. For AWG #14 wire, the DC resistance of the wire is 3.14 Ω/kft, as per Table 8, Chapter 9, NEC 2020. The voltage drop would be

$$V_{drop} = (1.25 \times 9.57) \times 3.14 \times 0.06 = 2.254 \, V,$$

which is less than 3.971 V. So, the cable sizing is correct.

6.3.6.2 Inverter to AC Distribution Panel

For the output current of the inverter, $I_{max} = 14.5$ A. So, the wire from the inverter should have a minimum ampacity of 14.5 × 1.25 = 18.125 A. Ignoring the temperature effect, at 75 °C, the required wire size from Table 310.16 of NEC 2020 is AWG #14 Copper THHW wire, which can carry up to 20 A of current.

Considering an allowable voltage drop of 1.5% in the inverter wire, the maximum voltage drop permitted is 220 V × 0.015 = 3.3 V.

Say the inverter is placed 10 ft away from the battery bank. So, the distance considered in this calculation will be 10 × 2 = 20 ft = 0.02 kft. For AWG #14 wire, the AC resistance of the wire is 3.1 Ω/kft, as per Table 9, Chapter 9, NEC 2020. So, the voltage drop in the wire will be

$$V_{drop} = 18.125 \times 3.1 \times 0.02 = 1.124 \, V,$$

which is less than 3.3 V. So, the cable sizing of #14 AWG copper uncoated wire is correct.

6.4   Distributed Solar PV Systems

Distributed PV systems are off-grid systems that are used for a dedicated purpose, such as driving an irrigation pump, lighting a street light, air quality measurement, powering a brooder house, outdoor aquarium, etc. One example of a distributed PV system as a PV-powered meteorological (MET) station is shown in Fig. 6.4. Two

Fig. 6.4 180 W solar PV-powered meteorological (MET) station at the Oregon Institute of Technology. This system includes a 12 V, 205 Ah deep-cycle AGM Battery

examples of distributed solar PV systems are explained in this chapter: solar PV-powered water pumping system and solar PV-powered street lighting system.

6.4.1   Water Pumping

Solar PV systems have diverse applications in agriculture, particularly for PV-fed water pumping systems, as shown in Fig. 6.5. For irrigation pumps, one may choose either an AC induction motor or a brushless DC (BLDC) motor. The motors can be powered by a solar PV system instead of consuming power from the utility grid. In this section, we will know about both these types of motors, study their comparative assessment, and go through three case studies related to irrigation pumps in three different places of the world.

Fig. 6.5 A 9.9 kW PV-fed water pump system in Sierra Cascade Nursery, Bonanza, Oregon, USA

6.4.1.1 AC Induction Motor

Since its invention, both single- and three-phase induction motors were widely used everywhere, from small residential applications, such as washing machines, up to huge industrial loads. Submersible water pumping was one of these applications in which induction motors were preferable over brushed DC motors that require the replacement of brushes and commutators. Brushes and commutators wear out due to their physical contact with the shaft and their low efficiency of 75–80%, whereas the efficiency of induction motors ranges from 85 to 97%. Moreover, induction motors are considered very reliable, maintenance free, and robust technology.

The rapid development in control, sensors, and power electronic fields led to new ways to control AC induction motors to enhance speed control, decrease surge current, and lower the power rating. Direct torque control, vector control, and VFD were dominant in most fields based on the application characteristics and requirements. VFD is the more comprehensive control technique that requires the DC current, the DC voltage, and the DC bus voltage to be sensed. In other techniques, current sensors are required for implementation. Thus, VFD is widely used in submersible water pumping for grid-fed and solar PV-fed water pumps.

Different factors were behind the late integration of PV into submersible water pumping. For instance, the necessity of designing the solar array based on the surge

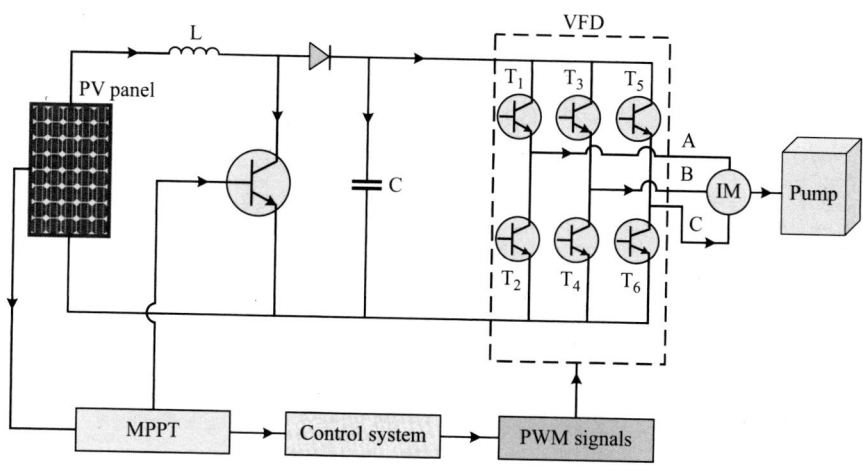

Fig. 6.6   Schematic diagram of a VFD-driven AC motor pump [7]

current is equal to 5 to 7 times the nominal operating current, which means a very high initial cost. On the contrary, VFD technology can eliminate the surge current and start the motor smoothly and incrementally until reaching the nominal speed without changing the motor shaft torque, which should be constant.

Figure 6.6 illustrates the AC induction motor pump configuration driven by VFD using the MPPT technique. First, the PV current and voltage are measured and sent to the MPPT module to apply one of the techniques discussed in Sect. 5.4.7.1. By using the MPPT technique, the operation point is located in the I–V curve. After determining the operating point, the MPPT algorithm controls the DC–DC converter to work as a buck or a boost converter to work in the MPP.

The final operating point is sent to a boost converter with a fixed output value to be applied in the input of the VFD inverter. Moreover, VFD is driven by PWM signals sent to the gates of the switches, which are mostly IGBTs[1] or BJTs.[2] The duration of the pulses (duty cycle value) and the order of applying the signals determine the voltage and frequency, where their ratio should be constant to keep the flux magnitude constant. The magnitude of the flux can be found by using the following equation [8]:

$$\psi = \frac{V}{\omega},\tag{6.9}$$

where

ψ = magnitude of flux, V = voltage, and ω = angular frequency.

---

[1] Insulated Gate Bipolar Transistors.

[2] Bipolar Junction Transistors.

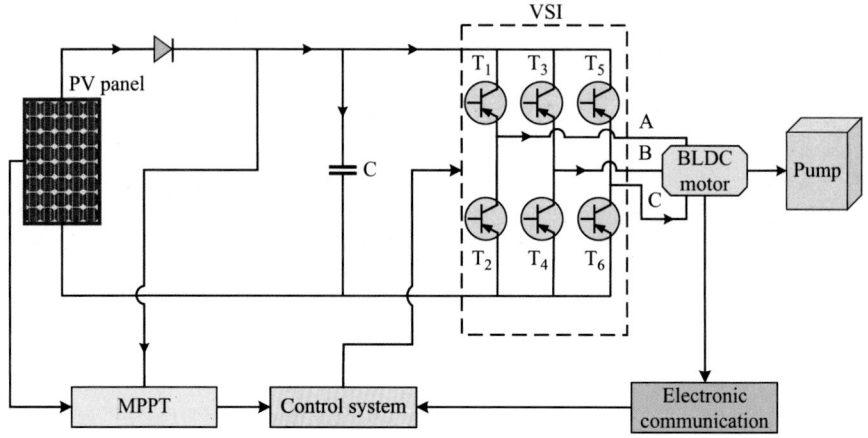

Fig. 6.7 Schematic diagram of a PV-fed BLDC motor pump [9]

6.4.1.2 Brushless DC Motor

AC induction motors are asynchronous machines, which means that the rotating field generated by the stator is faster than the actual spinning speed of the rotor. On the contrary, brushless DC (BLDC) motors (Fig. 6.7) are considered synchronous motors, where the rotating field should be in phase (synchronized) with the position of the turning field of the rotor. Moreover, AC motors cause significant power losses because they create the magnetic field by applying current to the stator windings, since $P = I^2 R$. But BLDC motors do not cause power losses in this way because the magnetic field in this technology is created by a permanent magnet material from which the rotor is manufactured. Therefore, BLDC motors reduce the ohmic losses.

BLDC motor is preferred for this water pumping application over the induction motor based on its characteristics (Fig. 6.9), such as high efficiency under different load factors, high reliability, low radio frequency interference, low maintenance, and low noise. The control system layout is shown in Fig. 6.8.

In order to operate the BLDC motor, the rotor position should be detected by photoelectrons, which generate Hall effect signals. The Hall effect refers to the production of a voltage difference across a conductor due to it being perpendicular to an electric field and both of them being perpendicular to a magnetic field. The Hall effect signals are sent to the controller to generate PWM signals to power the voltage source inverter (VSI), having IGBTs create the right pulse at the right time in the stator windings. Since the rotor is a permanent magnet, the coil formation in the stator should be energized by an electrical current. It is done to apply the simple principle of opposite pole attraction, where each phase of the coils is energized after each 60° zone to attract the rotor pole. Three Hall sensors (H1, H2, and H3) generate signals according to the position of the rotor to know which transistor should be switched on or off by applying or removing a gate signal to the IGBT of the VSI. The signal is shown in Table 6.8, where θ represents the angle of the rotor.

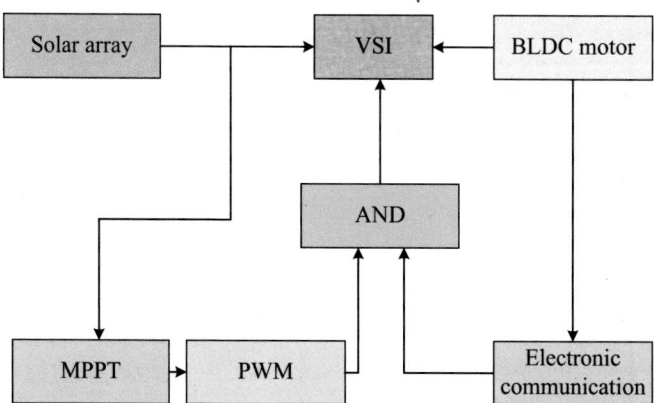

Fig. 6.8 Control system layout

Table 6.8 Signal generated by IGBT gate at different rotor angle and Hall effect signals [9]

Rotor angle, °	Hall signals			Signal of the IGBT's gate					
θ	H1	H2	H3	T1	T2	T3	T4	T5	T6
0–60	1	0	1	0	1	1	0	0	0
60–120	0	0	1	0	1	0	0	1	0
120–180	0	1	1	0	0	0	1	1	0
180–240	0	1	0	1	0	0	1	0	0
240–300	1	1	0	1	0	0	0	0	1
300–360	1	0	0	0	0	1	0	0	1

6.4.1.3 Comparison between BLDC- and VFD-controlled AC Induction Motor

Lorentz, the leading manufacturer in the solar water pumping industry, has studied the characteristics of both BLDC and AC motors to select the highest power rating of the system from which AC motors are more efficient than the BLDC ones. The company realized that it is preferable to use BLDC technology with a system with less than 7 kW solar modules capacity, mainly used to power not more than a 4 kW pump. This is because the dirt, temperature, and cable losses can easily reach 25%. In addition, solar irradiance may not reach 1000 W/m^2 in some geographical locations.

Generally, AC motors operate at their highest efficiency when they are connected at their rated load. However, for a motor fed by solar PV power, since the solar irradiance is the maximum for about 3–4 h only, the motor cannot operate at its rated load all the time and so has low efficiency. In contrast, a BLDC motor has a higher efficiency when the load factor varies from 0 to 40%. Moreover, it almost works at its maximum efficiency when the load factor is greater than 40%. Figure 6.9 demonstrates the comparison of the efficiency of the BLDC- and VFD-driven AC motors.

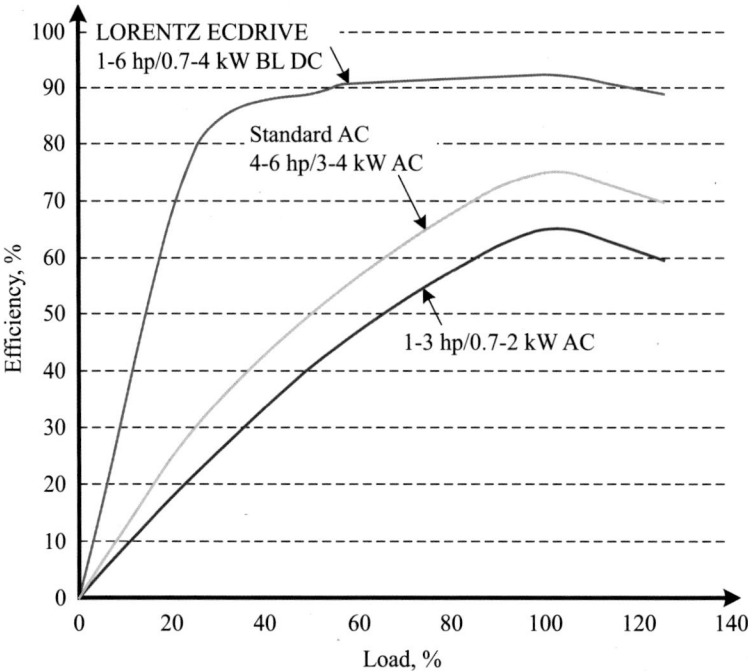

Fig. 6.9 Comparison of the efficiency of BLDC- and VFD-driven AC motors [10]

6.4.1.4 System Sizing Methodology

In order to start the calculation process correctly, the solar system designer should obtain some critical information from the landowner or by doing experiments and calculations. Thus, in this section, a detailed designing procedure will be conducted to determine all critical parameters required to select the sufficient PV modules and pump that satisfy the end-user's water requirement.

1. **Total Dynamic Head (TDH)**: TDH is the total equivalent vertical height that the water is expected to be pumped from the water surface in the submersible well into the highest point of the storage tank. The TDH should be determined based on the worst-case scenario (usually in summer) when the water level goes deeper than usual due to several factors, such as an increased usage and the absence of water compensation done by rain or springs. Moreover, the fraction of the pipeline should be considered by converting water friction pressure losses into its vertical head equivalent. The value of TDH can be found by using the following equation:

$$TDH = H_s + H_d + H_f, \tag{6.10}$$

where

Table 6.9 Daily average water requirements of users

Required water volume (L/day)	Water consumer
285	Human use (cooking, cleaning, and drinking)
151	Cows
75	Horses
8	Ships
15	100 chickens
80	Old trees
60	Young trees

H_s is the static head—the height from the underground to the ground level through which the pump has to suck water,

H_d is the discharge head—the height through which the pump sends the water from ground level to the tank, and

H_f is the friction equivalent head—the additional head due to the fittings and pipes.

The horizontal distance of the pipes, which are usually used between the surface of the well and the tank, can be added to the elevation head by substituting every 10 m of the horizontal pipes with 1 m of the vertical pipeline.

2. **Daily water requirement**: The daily water requirement for different situations is shown in Table 6.9.

3. **PSH on a tilted surface**: Since solar irradiance changes during the day and has different values along the year, it is important to know the equivalent period in which the pump will receive maximum power from the solar modules. This period can be defined as peak sun hours or PSH, which has been discussed in Sect. 4.11. If the average GHI/GTI of a month is known, the value of the average PSH can be found by using Eq. 4.5.

4. **Pump selection**: The total daily required water volume, TDH, and PSH values are the fundamental values to select the suitable model of a pump. One of the pump's curves should operate in the TDH and have the water flow that satisfies the following equation:

$$Pump\ flow\ rate \geq \frac{Total\ daily\ required\ water\ volume}{PSH}. \tag{6.11}$$

5. **PV array**: The submersible pump located in the well should receive sufficient power from the solar array to work properly and deliver the required water flow. However, PV modules face many losses based on the operating environment where the PV system is installed. Dirt losses (soiling factor) and high temperature are the most relevant factors to be addressed in any solar water pumping system design. Dirt losses can reach up to 10%, and temperature losses can reach up to 25%. Therefore, the power rating of the PV array can be obtained from the

following equation:

$$PV\ array\ power\ rating = \frac{Peak\ power}{\eta_{module} \times \eta_{thermal} \times \eta_{dirt}}. \qquad (6.12)$$

In deep submersible wells, the voltage drop due to ohmic loss of the long cable should be considered to avoid any failure in operation. Because when the pump's nominal voltage is greater than the received voltage, current flow from the source to the pump will increase. The series and parallel configuration of the PV modules is determined by the pump and controller manufacturers, who supply the designer with the recommended values of voltage and current that the system should operate in.

6.4.1.5 Case Studies
In this section, three case studies will be conducted to check the feasibility of solar water pumping applications by comparing them with other traditional power sources, such as diesel and the main utility grid. Since most arable lands are outside utility areas, diesel/gasoline electrical generators are the most common sources to run submersible water pumps. Three different continents are considered here for the case studies: Syria in Asia, Egypt in Africa, and Colombia in South America. These three locations are geographically diverse in terms of their climate and distance from the equatorial region. Syria is the farthest from the equator, followed by Egypt and then Colombia. They are all in the North hemisphere, and so their summer and winter times overlap; however, the driest month differs in these places, which is why the time for the greatest need for solar water pumping varies.

Case Study 1: Syria
Wells are the primary water source of Syrian agriculture due to their summer-grown crops and winter rainfall. This section aims to provide evidence of PV technology's efficiency based on a technical and financial comparison of agricultural water pumping techniques between PV and diesel. A Syrian farmer owns $12{,}000\,\mathrm{m^2}$ of land planted with 300 fruit trees, such as apples, pears, and pomegranates. Besides trees, he has livestock, such as fifteen sheep, six cows, a horse, and few chickens. All these plants and animals are watered by an AC pump installed on a submersible well. The depth of the well is 80 m, and the total dynamic head (TDH) is 40 m. The farmer used a diesel electricity generator since the utility grid does not reach his land. The study will discuss replacing the traditional system with a new solar PV system in addition to a new brushless direct current (BLDC) submersible pump.

Use of AC Submersible Pump
The traditional option is to keep using the same AC submersible pump. However, the power rating of the pump is 1125 W, and the water output is $1.5\,\mathrm{m^3/h}$, and it is powered by the 5 kVA diesel generator that consumes 1.5 L of diesel per hour. The farmer usually runs the generator according to the schedule shown in Table 6.10. The estimated cost of this option is described as the following:

Table 6.10 Pumping schedule for the small farm in Syria

Month(s)	Operating hours per day	Combined diesel consumption (m^3)
May, Oct, and Nov	8	1080
June	12	540
July, Aug, and Sept	15	2025
Rest of the year	4.5	1000
Total		4645

1. **Initial Cost:** The initial cost is considered $0 in this example since the farmer already had the pump and the generator.
2. **Operation and maintenance (O&M) cost:** The O&M cost is $2424, which can be broken down as:

   - Annual oil and filter change with maintenance = $102.
   - Annual diesel cost = $2322.

The total annual O&M cost is $2424 to extract 4645 m³ of water. The life cycle period of this system is estimated to be 20 years. So each year, the farmer has to pay a sum of $2424 until a new system is needed. The estimated cost of the AC pump and diesel generator is $1500.

Use of Solar PV-fed BLDC Submersible Pump
The new option is to use a solar PV-fed BLDC-driven submersible pump. The selection of the pump is based on the critical months with the highest energy demand. The critical months are July, August, and September, when the diesel generator runs for 15 h/day to pump 1.5 m³/h × 15 h = 22.5 m³/day. The PSH values at the optimal tilted angle (33°) for the months July, August, and September are shown in Table 6.11. The PSH in September is the lowest of these three months, so we will consider PSH = 7.02 in this case study.

In Fig. 6.10, the values of Global Tilted Irradiance (GTI) were found from the Global Horizontal Irradiance (GHI) values for Damascus, Syria (33.5138° N, 36.2765° E) using a MATLAB code, which is given in Sect. 4.16. The values show that the best tilt angle is the latitude angle and taken as 33 °C. Choosing the wrong angle might cause losses up to 6.7% in some cases, which might conclude in a failure in some applications, especially if the system has no oversizing margin.

The GTI is measured in kWh/m² per period. If GTI is calculated per day, this value will be equal to PSH. For further calculations, we take the GTI value from Fig. 6.10 and use it as the value of PSH.

The performance of the solar modules depends on the temperature of the solar cell, $T_{solar\ cell}$, along with the ambient temperature, $T_{ambient}$. The value of $T_{solar\ cell}$ can be found by using Eq. 2.6 as already shown in Sect. 2.14. For the study, the value of NOCT is taken as 46 °C as preferred by most manufacturers, and the irradiance is considered as 80 mW/cm². For example, for 12 °C ambient temperature, the temperature of the solar cell is

Table 6.11 Output water volume comparison of the AC submersible pump and the PV-fed BLDC motor pump in the farm in Syria

Month	Ambient temperature, °C	Solar cell temperature, °C	PSH at 33°	Power temp. coefficient, %/°C	Operating power, %	Water volume (AC), m³	Peak Power, W	Flow rate, m³/h	Water volume (PV-fed), m³	Difference, m³
Jan	12	38	4.9	−0.39	94.93	6.75	718	3.75	18.38	11.63
Feb	14	40	5.3		94.15	6.75	712	3.75	19.88	13.13
Mar	18	44	6.25		92.59	6.75	700	3.75	23.44	16.69
Apr	23	49	6.43		90.64	6.75	685	3.7	23.79	17.04
May	29	55	6.9		88.3	12	668	3.6	24.84	12.84
June	34	60	7.3		86.35	18	653	3.5	25.55	7.55
July	36	62	7.4		85.57	22.5	647	3.5	25.9	3.4
Aug	36	62	7.2		85.57	22.5	647	3.5	25.2	2.7
Sep	33	59	7.02		86.74	22.5	656	3.6	25.27	2.77
Oct	27	53	6.6		89.08	12	673	3.6	23.76	11.76
Nov	21	47	6.4		91.42	12	691	3.7	23.68	11.68
Dec	14	40	4.7		92.59	6.75	700	3.75	16.5	9.75

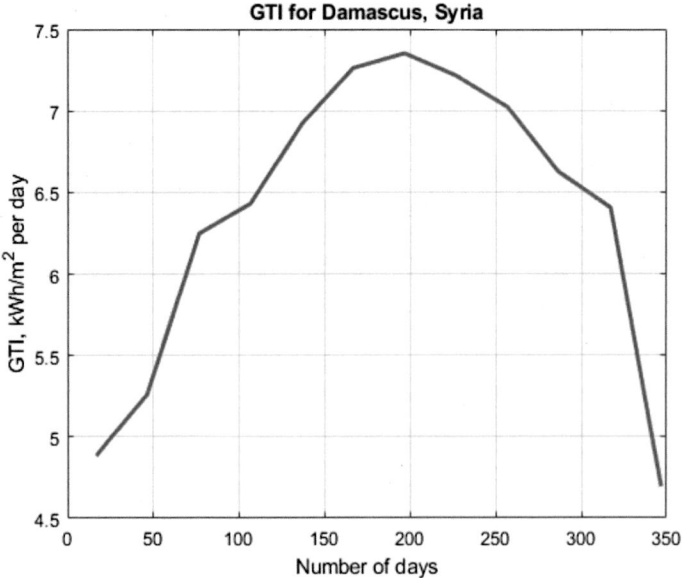

Fig. 6.10 The GTI for Damascus, Syria, at a 33° tilt as found from the MATLAB program

$$T_{solar\ cell} = T_{ambient} + \frac{NOCT - 20}{80} \times S = 12 + \frac{46 - 20}{80} \times 80 = 38\,°C.$$

Likewise, the solar cell temperature can be determined for any value of ambient temperature, provided that the NOCT and irradiance values are known. These temperatures also affect the operating voltage, current, and power. The solar cell temperatures at different ambient temperatures are calculated in the above manner and enlisted in Table 6.11. The cell temperature can also be used to find the operating power, considering the effect of temperature using Eq. 3.7. For example, the operating power considering the temperature loss for a solar cell temperature of 38 °C is determined as shown in the following:

$$Operating\ power\ considering\ temperature,\ \% = P_{TC} \times (T_{cell} - 25) + 100$$
$$= (-0.39) \times (38 - 25) + 100 = 94.93\%,$$

where P_{TC} is the power temperature coefficient, $\%/°C$.

Among the three peak months, September has the worst irradiance. So, we need to use the PSG value for September in this calculation. Now, the flow rate of the pump can be determined by the following equation:

$$Flow\ rate\ of\ pump = \frac{Total\ daily\ required\ water\ volume}{PSH}$$
$$= \frac{22.5}{7.02} = 3.205\,\text{m}^3/\text{h}.$$

So, the required pump should have a minimum flow rate of 3.205 m³/h to satisfy the demand. From Fig. 6.11, the curves show that the pump can pump 3.5 m³/h of water if it receives about 650 W from the PV modules, regardless of losses that occur by dirt, temperature, and degradation with time. The value of peak power is selected from the curve, which is for the TDH value of 40 m. Thus, the power rating of the array in the 20th year of operation is determined as follows:

$$PV\ array\ power\ rating = \frac{Peak\ power}{\eta_{thermal} \times \eta_{dirt} \times \eta_{module}}$$
$$= \frac{650}{0.86 \times 0.90 \times 0.80} = 1050\,W,$$

where

- The power for temperature loss at the maximum ambient temperature of the critical month (September), $\eta_{thermal} = 86\%$ (approximately) $= 0.86$. So, the temperature loss is $(100–86)\% = 14\%$.

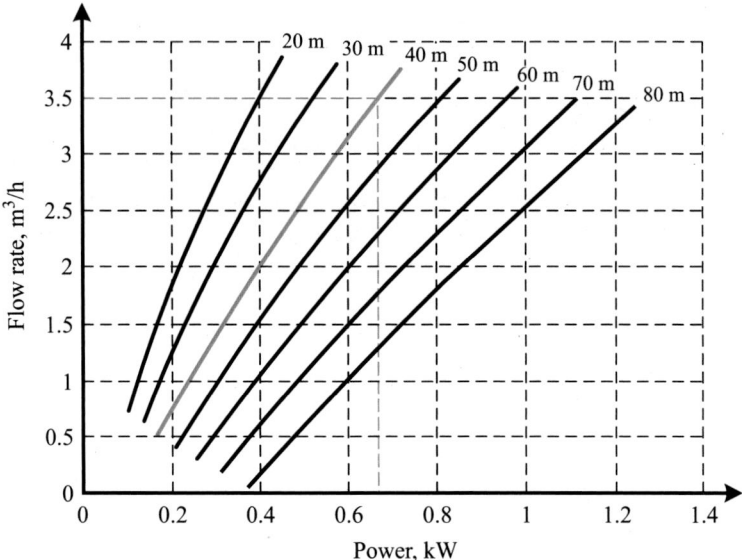

Fig. 6.11 Flow rate and power characteristics of the solar pump in Syria

- Dirt losses are 10%. So, $\eta_{dirt} = 0.90$.
- The minimum module efficiency during the lifetime of the project is 80%. So, $\eta_{module} = 80\% = 0.80$.

By using Eq. 6.12, the peak power at different ambient temperatures can be calculated, as the power rating of the PV array is now known. The output water volume for using both traditional AC pump and solar fed pump is shown in Table 6.11. The output water volume for the AC submersible pump is calculated by multiplying water output 1.5 m³/h with the operating hours as shown in Table 6.10. On the contrary, the output water volume for solar PV-fed BLDC submersible pump is calculated using Eq. 6.11.

The selection criteria for the solar PV-fed BLDC submersible pump was conducted based on a worst-case scenario. The designer might adjust the efficiencies based on several factors, such as the required project lifetime, the awareness of the farmer, and the surrounding atmosphere of the system. These factors determine the dirt losses, which can be neglected if the solar module undergoes a periodic cleaning process. So, the PV system designer or installer may recommend the farmer to conduct the cleaning operations when the modules are not exposed to high irradiance and temperatures. The solar cell temperatures might reach more than 90 °C. So, adding cold water to such a hot glass layer of the modules can apply undesired stress on the glass, which might end up forming a crack.

The output comparison between the mentioned two ways of powering the pump is shown in Fig. 6.12. It can be seen that the water pumped from the solar PV-fed

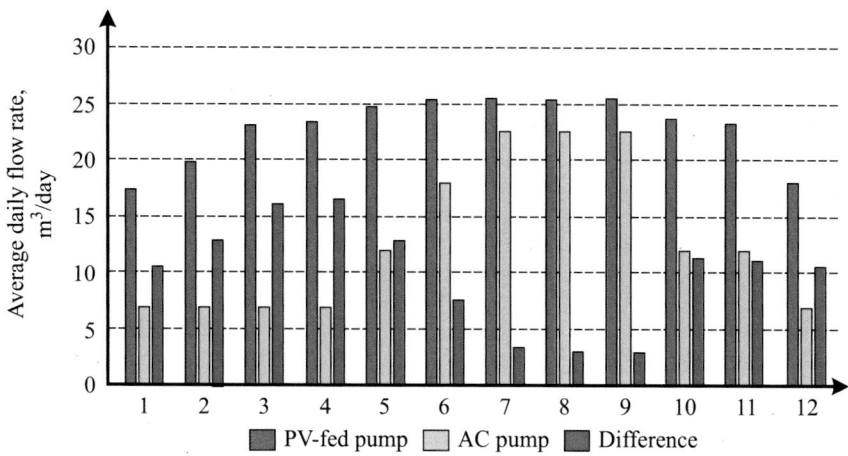

Fig. 6.12 Output comparison between the PV-fed pump and the AC pumps in Syria

pump is more than the water found by using the traditional AC pump. So, using the solar PV-fed BLDC motor pump is more preferable to using the AC pump in terms of efficiency. Besides, the farmer can install the solar PV-fed pump by investing $4000 initially, which will fulfill the water requirements of the farm.

Case Study 2: Egypt

The second case study is in Aswan, Egypt (24° N, 32° E), where the depth of the well is 80 m and the TDH is 45 m. The daily average water demand, in this case, is constant along the months of the year (unlike Case-1). The land–water requirement is satisfied by filling a 40 m^3 water reservoir two times a day. So, the submersible water pump is supposed to supply the tank with $40 \times 2 = 80$ m^3 every day.

The solar resources of the selected location can be observed from Fig. 6.13, which shows the variation of GTI at a 24° tilt angle. Similar to the previous case, the value of PSH is taken from Fig. 6.13, as GTI is considered to be equal to PSH if the period is 1 day.

For the study, the value of NOCT is taken as 46 °C as preferred by most manufacturers, and the irradiance is considered as 80 mW/cm^2. For example, for 25 °C ambient temperature,

$$T_{solar\ cell} = T_{ambient} + \frac{NOCT - 20}{80} \times S = 25 + \frac{46 - 20}{80} \times 80 = 51\ °C.$$

Using Eq. 3.7, the operating power considering the effect of temperature loss at different ambient temperatures is calculated as shown in Table 6.12.

Here,

$$Flow\ rate\ of\ the\ pump = \frac{Total\ daily\ required\ water\ volume}{PSH}$$

$$= \frac{80}{6.1} = 13.1\ m^3/h.$$

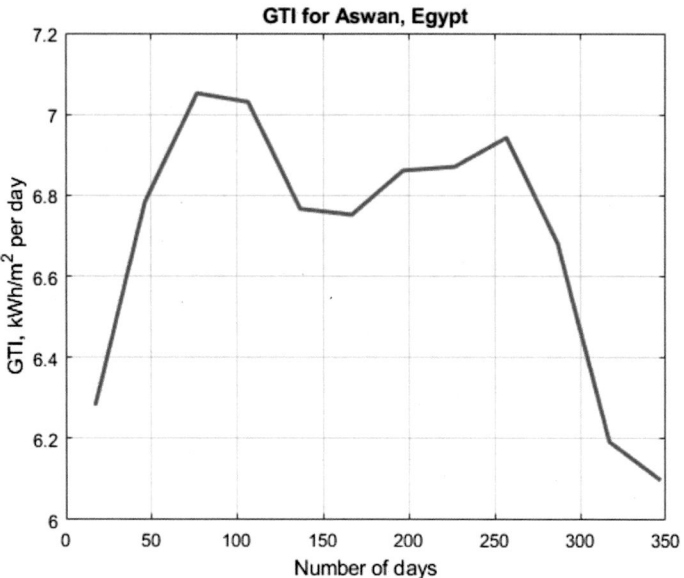

Fig. 6.13 The GTI for Aswan, Egypt, at a 24° tilt as found from the MATLAB program

Table 6.12 Monthly water output of solar PV-fed BLDC pump in Egypt

Month	Ambient temperature, °C	Solar cell temperature, °C	PSH at 24°	Power temp. coefficient, %/°C	Operating power, %	Peak power, W	Flow rate, m³/h	Water volume (PV-fed), m³
Jan	23	49	6.3	−0.39	90.64	2669	13.5	85.05
Feb	25	51	6.8		89.86	2646	13.5	91.8
Mar	30	56	7.09		87.91	2589	13.4	95.006
Apr	35	61	7.05		85.96	2532	12.2	86.01
May	40	66	6.78		84.01	2474	11.8	80.004
June	40	66	6.76		85.57	2520	12	81.12
July	40	66	6.87		85.57	2520	12	82.44
Aug	40	66	6.88		85.57	2520	12	82.56
Sep	40	66	6.9		85.57	2520	12	82.8
Oct	36	62	6.68		85.57	2520	12	80.16
Nov	30	56	6.19		87.91	2589	13.4	82.946
Dec	25	51	6.1		89.86	2646	13.6	82.96

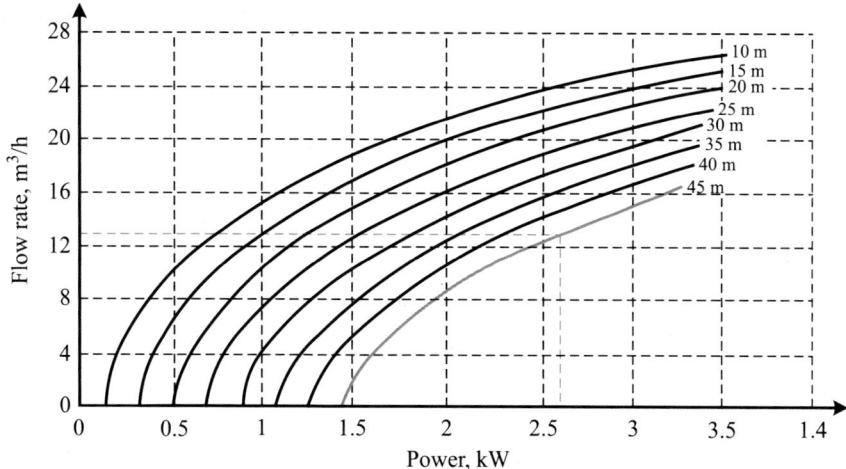

Fig. 6.14 Flow rate and power characteristics of the pump in Egypt

As per the calculations, the required pump should have a minimum flow rate of 13.1 m³/h. From Fig. 6.14, the curves show that the chosen pump at TDH of 45 m can pump 13.5 m³/h of water if it receives 2650 W from the PV modules, regardless of losses that occur by dirt, temperature, and degradation with time. The value of peak power is selected from the curve, which is for the TDH value of 45 m. Thus, the power rating of the array is determined as follows:

$$PV\ array\ power\ rating = \frac{Peak\ power}{\eta_{thermal} \times \eta_{dirt} \times \eta_{module}}$$

$$= \frac{2650}{0.90 \times 0.95 \times 0.80} = 3875\ W,$$

where

- The power for temperature loss at the maximum ambient temperature, $\eta_{thermal} = 90\%$ (approximately) $= 0.90$. So, the temperature loss is $(100-90)\% = 10\%$. The losses were calculated based on the highest temperature in December (25 °C).
- Dirt losses are 5%. So, $\eta_{dirt} = 0.95$.
- The minimum modules' efficiency during the lifetime of the project is 80%. So, $\eta_{module} = 0.80$.

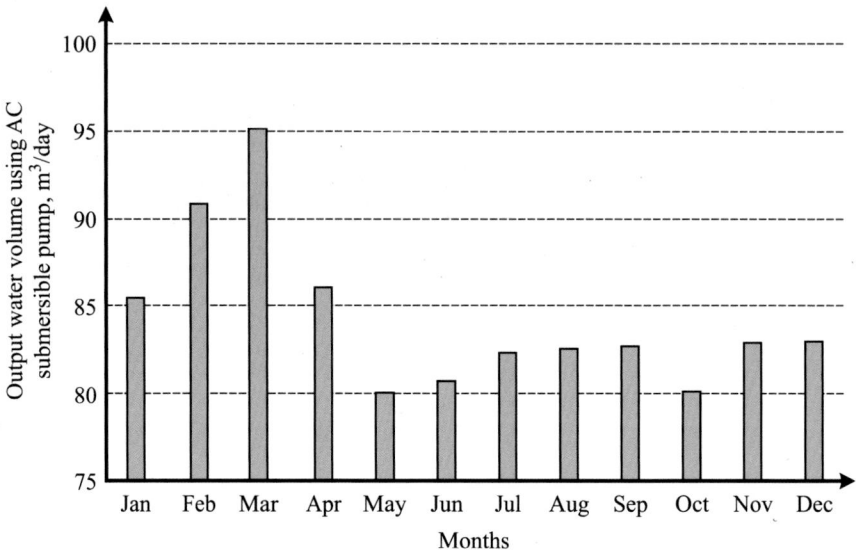

Fig. 6.15 Monthly water output of the pump in Egypt

By using Eq. 6.12, the peak power at different ambient temperatures can be calculated, as the power rating of the PV array is now known. The output water volume for using a solar PV-fed pump is shown in Table 6.12. The output water volume for the solar PV-fed BLDC submersible pump is calculated using Eq. 6.11 at different temperatures. Moreover, the output of water from the pump in Egypt in a year is shown in Fig. 6.15.

Case Study 3: Colombia

The third case study is in Ovejas, Colombia (9.5° N, 75.2° W), which has a TDH of 90 m. The maximum daily average water demand, in this case, is 165 m^3 per day in January, while the rest of the year, it is not more than 100 m^3 per day. The solar resources of the selected location can be observed from Fig. 6.16, which shows the variation of GTI at a 9° tilt angle. Similar to the previous two cases, the value of PSH is taken from Fig. 6.16 as the GTI is considered to be equal to PSH if the period is 1 day.

For this study as well, the value of NOCT is taken as 46 °C as preferred by most manufacturers, and the irradiance is considered as 80 mW/cm^2. For example, for 30 °C ambient temperature,

$$T_{solar\ cell} = T_{ambient} + \frac{NOCT - 20}{80} \times S = 30 + \frac{46 - 20}{80} \times 80 = 56\,°C.$$

Using Eq. 3.7, the operating power considering the effect of temperature at different ambient temperatures is calculated as shown in Table 6.13.

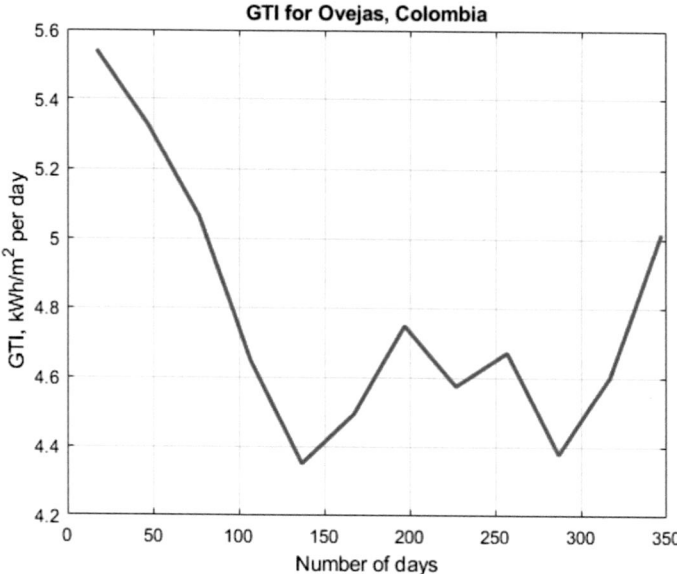

Fig. 6.16 The GTI for Ovejas, Colombia, at a 9° tilt as found from the MATLAB program

Table 6.13 Monthly water output of solar PV-fed BLDC pump in Colombia

Month	Ambient temperature, °C	Solar cell temperature, °C	PSH at 9°	Power temp. coefficient, %/°C	Operating power, %	Peak power, kW	Flow rate, m³/h	Water volume (PV-fed), m³
Jan	30	56	5.56	−0.39	87.91	13.086	29.8	165.688
Feb	33	59	5.35		86.74	12.912	29	155.15
Mar	33	59	5.08		86.74	12.912	29	147.32
Apr	33	59	4.66		86.74	12.912	29	135.14
May	33	59	4.36		86.74	12.912	29	126.44
June	33	59	4.5		86.74	12.912	29	130.5
July	33	59	4.75		86.74	12.912	29	137.75
Aug	33	59	4.58		86.74	12.912	29	132.82
Sep	32	58	4.67		87.13	12.97	29.2	1136.364
Oct	30	56	4.38		87.91	13.086	29.8	130.524
Nov	30	56	4.6		87.91	13.086	29.8	137.08
Dec	32	58	5.02		87.13	12.97	29.2	146.584

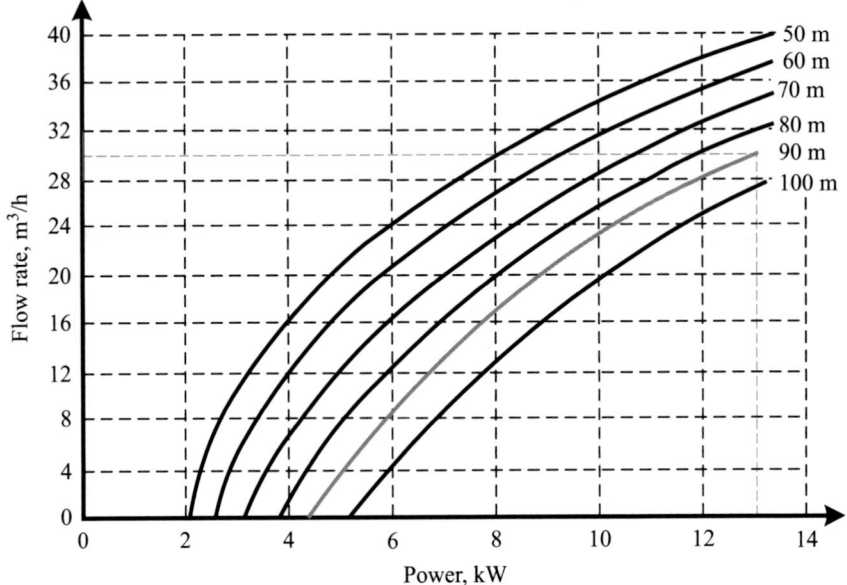

Fig. 6.17 Flow rate and power characteristics of the pump in Colombia

Here,

$$Flow\ rate\ of\ the\ pump = \frac{Total\ daily\ required\ water\ volume}{PSH}$$

$$= \frac{165}{5.56} = 29.68\,\text{m}^3/\text{h}.$$

As per the above calculation, the required pump should have a minimum flow rate of 29.68 m³/h. From Fig. 6.17, the curves show that the pump can pump 30 m³/h of water if it receives 13.1 kW from the PV modules, regardless of losses that occur by dirt, temperature, and degradation with time. The value of peak power is selected from the curve, which is for the TDH value of 90 m. Thus, the power rating of the array is determined as follows:

$$PV\ array\ power\ rating = \frac{Peak\ power}{\eta_{thermal} \times \eta_{dirt} \times \eta_{module}}$$

$$= \frac{13,100}{0.88 \times 0.95 \times 0.80} = 19.587\,\text{kW},$$

where

- Temperature losses in January reaches 12%. The losses were calculated based on the highest temperature in January (30 °C) for $\eta_{thermal} = 88\%$ (approximately) $= 0.88$.

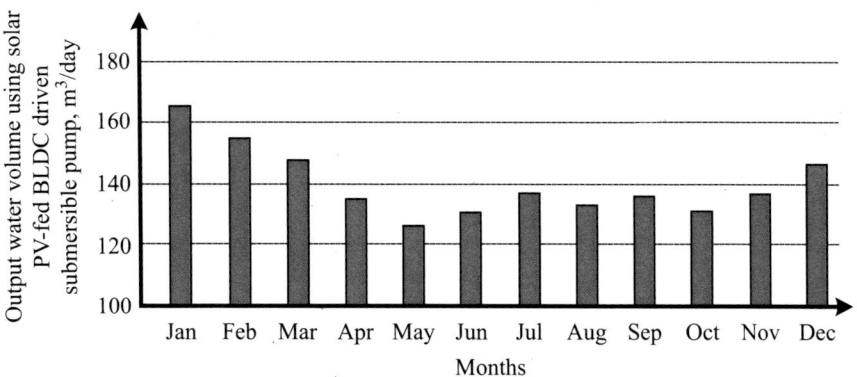

Fig. 6.18 Monthly water output of the pump in Colombia

- Dirt losses are 5%. So, $\eta_{dirt} = 0.95$.
- The minimum module efficiency during the lifetime of the project is 80%. So, $\eta_{module} = 0.80$.

By using Eq. 6.12, the peak power at different ambient temperatures can be calculated, as the PV array's power rating is now known. The output water volume for using a solar PV-fed pump is shown in Table 6.13. The output water volume for the solar PV-fed BLDC submersible pump is calculated using Eq. 6.11 at different temperatures. Moreover, the output of water from the pump in Colombia in a year is shown in Fig. 6.18.

6.4.1.6 Effect of Cable Length

The chosen pump is operated by a VFD-driven three-phase induction motor. The pump and its motor are submerged at 110 m depth underground in the well, while the VFD controller is on the surface. So, the distance between the pump motor and the VFD controller is 110 m. Therefore, the length of the cable that transfers the current and the voltage signals to the motor is not less than 110 m in the case of installing the VFD drive exactly at the surface of the well. According to Article 430.122 of NEC 2020 [11], the cables in an adjustable speed drive system should be designed to stand 125% of the nominal output current and maintain the voltage drop down to 1–2%. Designers and installers should follow all the recommended steps provided by the national standards and the manufacturers regarding cable sizing, over-current protection devices, and appropriate pump, modules, and VFD selection. However, the system will most likely stop running within the first few days of operation.

To comprehend the reason, it is essential to understand that solar water pumping is considered an off-grid system. However, the load and its driving method require the installer to have a deep knowledge of electronics, motor types, and their driving methods and power theories. VFD drivers contain very high switching-frequency devices that change their mode between off and on in the range of 2–8 kHz and

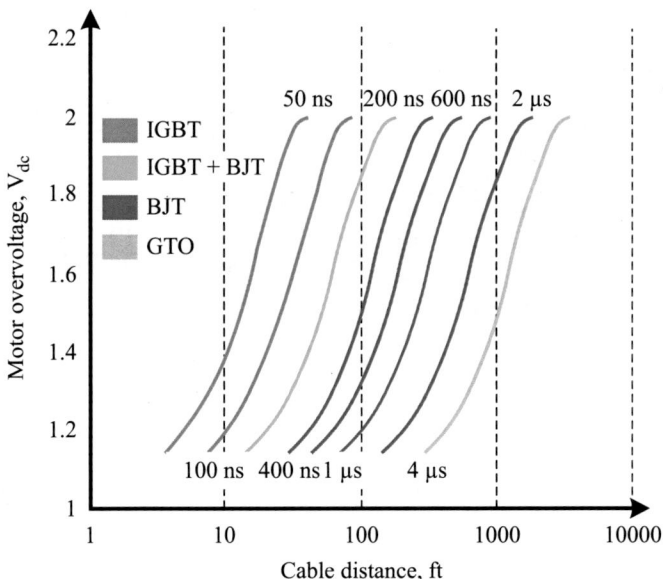

Fig. 6.19 Effect of cable length on motor overvoltage. Source: Series A 440–600 V_{AC} dv/dt manual (2004) from MTE Corporation [13]

sometimes 16 kHz. In other words, the status of the transistor changes from 62.5 to 500 microseconds to generate PWM signals that are similar but not exact to the pure sine wave that the motor is supposed to work at.

The high-frequency characteristic of the VFD has the disadvantage of generating very fast transient voltages. These voltages are transmitted to the motor and add extra electrical stress to its winding insulations. This disadvantage always appears in deep wells since the transient magnitude increases with the cable distances between the frequency controller (VFD) and the motor. Figure 6.19 illustrates the effect of different frequencies and cable length on the overvoltage applied to the motor, which leads to a failure in the system. Therefore, VFD manufacturers highly recommend using an electrical filter to mitigate this disadvantage [12].

Filters are of two types: dv/dt filters and sinewave filters. Both filters mainly contain inductors and capacitors to limit the voltage spikes and rapid voltage changes generated by VFD and the long motor cable. However, since wave filter components have higher power ratings, they smoothen the output signal of the VFD. Therefore, a shape closer to the pure sine wave is generated by the system. The selection process of the filter is based on the length of the motor cable and the nominal power rating of the system. The rule of thumb that most well-known manufacturers use is that filters are highly recommended to all systems with a power rating of more than 11 kW and a cable length greater than 50 m (164 ft). Furthermore, 150 m is the limit after which the sinewave filter is preferable over the dv/dt filter.

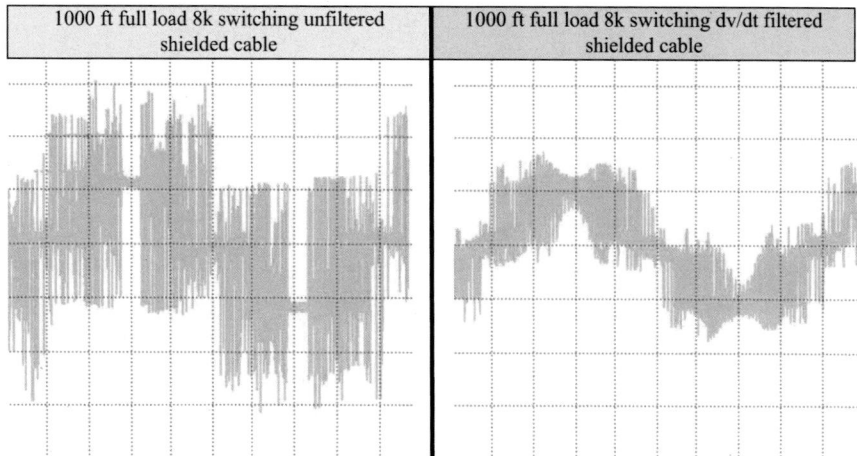

| 1000 ft full load 8k switching unfiltered shielded cable | 1000 ft full load 8k switching dv/dt filtered shielded cable |

Fig. 6.20 The noises in the current are reduced due to using the dv/dt filter [13]

In all filter type selection processes, the chosen filter should satisfy the following two factors:

1. It should handle 110% of the full load motor current [14].
2. It should be designed to handle the nominal switching frequency of the VFD inverter.

The effect of using the dv/dt filter can be observed from the juxtaposition of the two load curves in Fig. 6.20. Notice how the noises in the cable are reduced in the right graph because of using the dv/dt filter.

6.4.2   Street Lighting Systems

Street lighting systems, one shown in Fig. 6.21, have recently gained increased importance due to the rapid acceleration of technological development. It aims to maintain the highest comfort and safety standards with the lowest possible expenses balanced against the environmental impact. The International Commission on Illumination (CIE) defined the main purposes of public street lighting systems as the following:

1. Allow all road users, such as motorists, pedestrians, animal-drawn vehicle drivers, and pedal cyclists, to proceed safely.
2. Give the pedestrians the ability to detect hazards, recognize others, orient themselves, and most importantly, give them a sense of security.
3. Enhance human vision during both daytime and dark times.

Fig. 6.21 A solar PV-powered street lighting system. These systems are prevalent in many parts of the world

These points contain several latent purposes such as crime discouragement, decreased night-time fatal accidents, and the neighborhoods' sense of being inhabited and protected. As a matter of fact, these systems roughly accounted for 1–4% of the overall national energy consumption in Syria. Therefore, most cities started to do retrofitting plans to convert the conventional AC inefficient lamps with LEDs, which is broadly considered a financially and technologically feasible enterprise. The non-redundant structure of a standalone solar street lighting system contains the following:

- Solar module(s), responsible for providing the required night-time energy demand
- Storage system, to store the energy generated by the solar modules in the daytime and deliver it to the load at night in the absence of solar irradiance
- Charge controller, the "brain of the system" due to its ability to fully control the system by connecting, disconnecting, changing the brightness of the load,

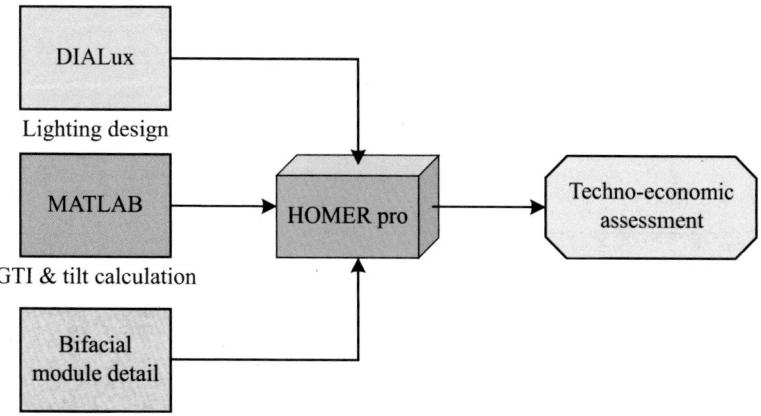

Fig. 6.22 General methodology of work for the street lighting system design

and controlling the solar modules' current to keep the storage system in a good operating condition
- The load, usually an LED lamp

6.4.2.1 Design Case Studies

In this section, the design of a street lighting system is going to be walked through. The street lighting system is relatively simpler than other systems having only one load. The advantages include neglected voltage drop, very simple protection calculations, and the mentioned system components can operate based on different technologies. For instance, batteries can be lithium-ion based or lead-acid based. However, it is essential to know that the most economical configuration of these technologies meets all lighting standards and has the shortest payback period for the initial and operating costs.

Therefore, in this section, we will study a published roadmap on designing a DC LED solar street lighting system based on British lighting standard EN 13201-2:2015. We will then technically and economically compare 12 different scenarios of energy storage (Li-ion and lead-acid batteries) fed by monofacial modules or bifacial modules with two different albedo values (20% and 40%) to feed two luminaires, which meet M3 and M4 lighting class standards. The lighting class M is set by the EN 13201-2:2015 standard, and class M specifies motorized traffic on routes with medium to high speeds. Furthermore, it will have a base case of a grid-fed system using well-known software used in the industry. The overall method of this design is roughly presented in Fig. 6.22.

A four-lane street was simulated in Dialux software to determine the optimum luminaire model that meets several lighting standards for a specific lighting class. The output of the Dialux software is input to HOMER PRO software to assist in conducting financial analysis and making a final decision about the most feasible option to be applied. MATLAB and System Advisor Model (SAM) are also used

Table 6.14 Road geometry for classes M3 and M4

Lighting class according to EN1 3201-2	M3 and M4
Road width, m	14
Number of lanes	4
Luminaire height, m	8
Overhang, m	0
Pole spacing, m	36

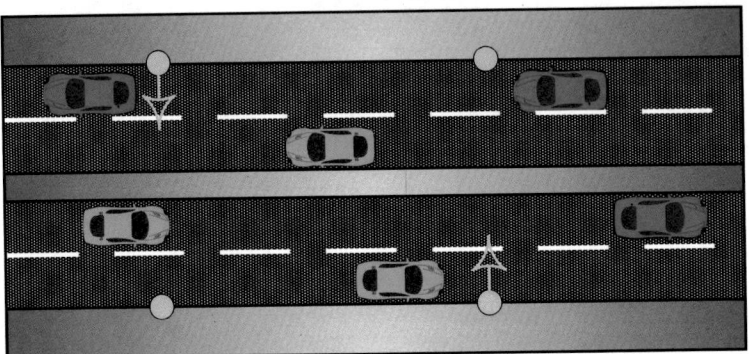

Fig. 6.23 The proposed case geometry with its lighting arrangement

in the design phase as inputs to HOMER PRO to obtain the best tilt angle for photovoltaic modules and to simulate the bifacial modules, respectively.

HOMER PRO is a software established by the National Renewable Energy Laboratory (NREL) and further developed by Dr. Peter Lilienthal. HOMER PRO simulates the operation process of any system (street light luminaire in this case study) by making energy balance calculations between the load and the energy sources (grid or PV and battery). However, it keeps the resolution down to 1 min to ensure that the system's resources can supply the load adequately. Moreover, the software algorithm provides the total present value of the initial and operating cost of the desired systems based on input inserted by the user, which is discussed in detail in this chapter. The economic methods through which the proposed scenarios are compared are the net present cost (NPC) and the levelized cost of electricity (LCOE). The system with the least LCOE and NPC is considered as the best-case scenario.

Dialux software was used in the selection process of the most appropriate luminaire that meets all the lighting class requirements with the lowest power rating. This results in the reduction of energy consumption, which minimizes solar PV modules and battery systems. Double two-lane streets with an opposite arrangement lighting of a carriageway where there are no adjacent areas with separate lighting requirements (M3 and M4) were simulated in the software based on the geometry listed in Table 6.14 and graphically represented in Fig. 6.23. Here, ENI 3201 is a road light model, standard of road lighting.

Table 6.15 Average daily and monthly lighting operation hours [15]

Month	Daily average lighting operation time	Total monthly lighting operation time (h)
January	13 h 47 min	427.28
February	13 h 0 min	377
March	12 h 0 min	372
April	10 h 57 min	328
May	10 h 5 min	312.58
June	9 h 39 min	289.5
July	9 h 53 min	306.38
August	10 h 39 min	330
September	11 h 38 min	349
October	12 h 40 min	392
November	13 h 34 min	407
December	14 h 1 min	434.5
Total		4325.24

The simulation results from Dialux software showed that luminaires with power ratings of 51 W and 39 W are required to meet M3 and M4 classes' standards, respectively. These power ratings are multiplied by the daily average lighting operation time acquired from the applied systems in a middle eastern country. It is done to obtain a 15-min interval load profile, as shown in Table 6.15.

Once the load profile is obtained, the next step is to study the energy source parameters to guarantee a sufficient supply for the load demand. Monofacial modules are tilted with the optimum tilt angle, obtained from the MATLAB code. It guarantees the maximum GTI in the critical month (December) when the luminaire has the maximum operating hours (14 h and 1 min). The steps of the methodology are shown in the flowchart of Fig. 6.24.

6.4.2.2 Bifacial Technology Inputs

The most recent HOMER PRO version cannot simulate bifacial modules. Therefore, another NREL software called SAM is used to simulate the bifacial gain based on the following cases:

- Two different albedo values (20% and 40%)
- Pole height of 8 m, which represents the height of the bifacial PV module

Solar data are imported from the recommended solar resource of the European Commission's Science and Knowledge Service. It is worth noting that SAM is designed for grid-connected applications. Thus, the additional gain by the rear side is obtained by comparing the two one-module systems (monofacial and bifacial) connected to an oversized inverter. The oversized inverter is used to avoid any clipping when the DC/AC ratio is greater than 1.

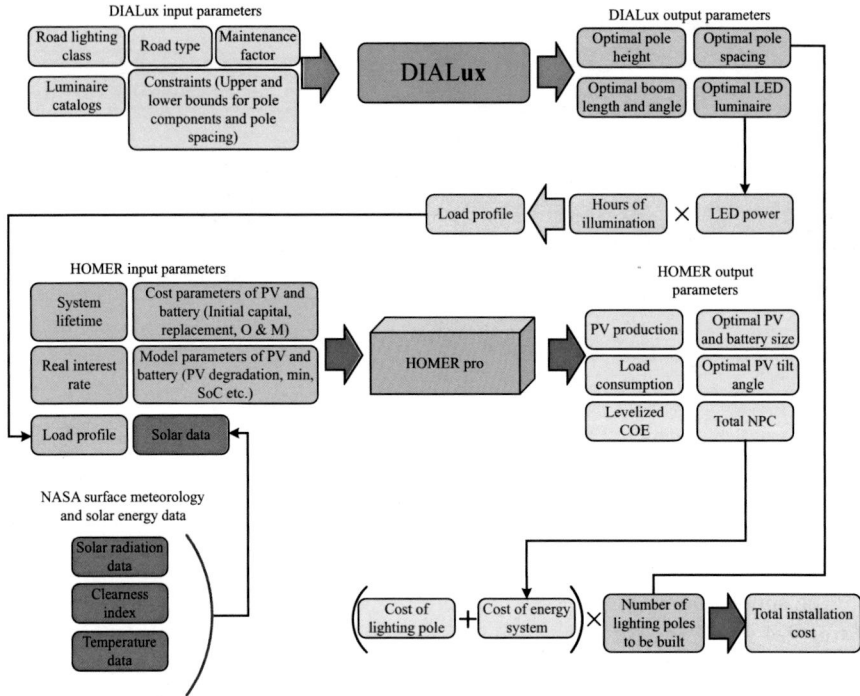

Fig. 6.24 Flowchart of the detailed steps of the design methodology

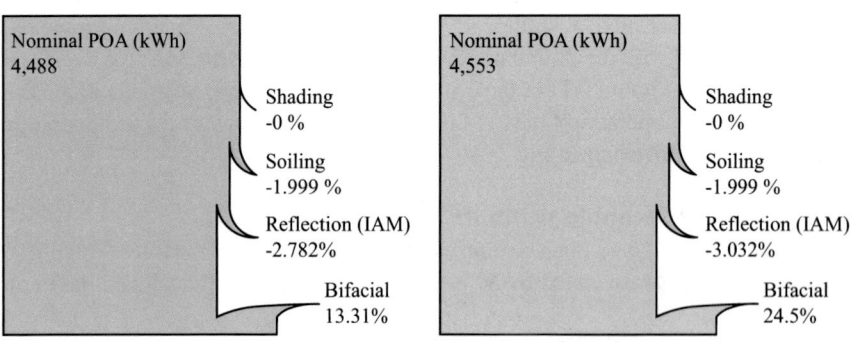

Fig. 6.25 Bifacial gain when albedo is 20% (left) and 40% (right)

Figure 6.25 illustrates the gain of the bifacial modules when installed on a pole and facing the south with two different albedos (20% and 40%). The bifacial gain was added to HOMER PRO as a derating factor with a value greater than 100% to simulate the extra energy yield over the project lifetime.

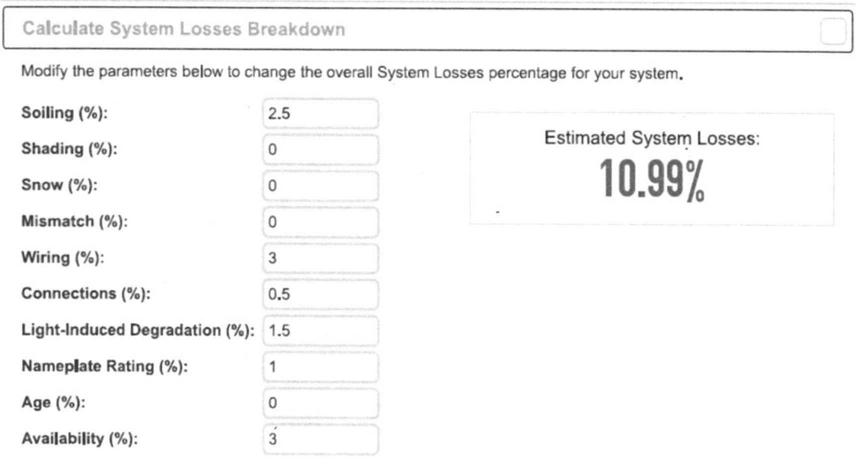

Calculate System Losses Breakdown

Modify the parameters below to change the overall System Losses percentage for your system.

Soiling (%):	2.5
Shading (%):	0
Snow (%):	0
Mismatch (%):	0
Wiring (%):	3
Connections (%):	0.5
Light-Induced Degradation (%):	1.5
Nameplate Rating (%):	1
Age (%):	0
Availability (%):	3

Estimated System Losses:

10.99%

Fig. 6.26 Estimated system losses

6.4.2.3 Selection of PV Module

The first energy source is the PV array, which has three different scenarios: monofacial, bifacial with 20% albedo, and bifacial with 40% albedo. Therefore, the PV module should have a bifacial version with the same characteristics to make the comparison more accurate. The PS-M72 (385 W) [16, 17] from Philadelphia Solar was selected and added to the HOMER Pro library using the data from its datasheet. The cost of 1 kW was $400, which also serves as a replacement cost after the lifetime. However, the replacement cost of the PV module is not necessary because the overall lifetime of the whole project is 25 years. In other words, it will not be considered in the calculations. The operating cost was estimated as $2 per standalone PV on each pole, which represents the cost for the periodic cleaning process.

The rating of PV modules is set based on standard conditions, i.e., neglecting all the losses caused for different factors such as high temperature, low irradiation, etc. So, in real-life applications, the output is less than the rated value. The derating factor (DF) determines how much less the output is going to be. The DF is a value that represents the reduction in the actual output power compared with the rated value. For example, if the DF of a PV array is 70%, then it implies that the array would produce 70% of the rated output in real-life applications. The DF value is found by using PVWatts, the recommended resource by NREL, as shown in Fig. 6.26. The DF of a bifacial module can be calculated using Eq. 6.13. For the three types of PV modules, i.e., monofacial, bifacial with 20% albedo, and bifacial with 40% albedo, the technical and financial input details are shown in Table 6.16.

$$DF_{bifacial} = DF_{monofacial} \times Bifacial\ gain_{SAM} \qquad (6.13)$$

Table 6.16 The technical and financial details of the PV module

Factors	Monofacial	Bifacial (Albedo 20%)	Bifacial (Albedo 40%)
Initial cost, $/k$W_p$	400	488	488
Replacement cost, $/k$W_p$	400	488	488
Operation and maintenance cost, $/year	2	2	2
Derating factor (DF)	89	100	110.80

Table 6.17 Energy storage technical and financial input details

Factors	Lead acid	Li-ion
Model	Trojan SAGM 12 105	Trojan Trillium 12.8-110
Technology	Deep-cycle solar AGM	Lithium iron phosphate
Nominal voltage, V	12	12.8
Nominal capacity, Ah	105	110
Nominal energy, kWh	1.26	1.4
Capital cost, $	290	1075
O&M cost $/year	0	0
Initial state of charge, %	100	100

6.4.2.4 Energy Storage

Two battery technologies are considered in this design: deep-cycle lead-acid and lithium iron phosphate batteries. The batteries of choice were from the Trojan Battery Manufacturer. The models were also added to the library after consulting the company to obtain the accurate data that are not available in their datasheets. Table 6.17 summarizes the main inputs in the energy storage section. The table shows the basic datasheet of the mentioned batteries. Once the lifetime of the batteries ends, they can be sold, which will reduce the replacement cost.

6.4.2.5 Advanced Grid

The basic case that all scenarios are compared with is the existing grid-fed LED streetlamp. The grid is modeled and simulated as a robust grid with 100% availability. The grid power price is set as 0.164 $/kWh.

6.4.2.6 Converter

The design case contains a 220 V, 50 Hz grid, a rectifier (220 V_{AC} to 12 V_{DC}), and an LED load. The HOMER PRO library does not include a pure rectifier but a converter that works as a rectifier (AC/DC) and an inverter (DC/AC) at the same time. To inactivate the inverter side of the converter, the inverter efficiency was set to 0.01%. In comparison, the rectifier efficiency was set as 93%, as provided from the datasheet of the rectifier used.

The capital cost for both M3 and M4 scenarios is $40, which is also the replacement cost after an expected 10-year lifetime of the power electronic-based rectifier.

6.4.2.7 Project Lifetime, Annual Capacity Shortage Inflation, and Discount Rate

Since this case study is considered a long-term enterprise, it is difficult to predict the economic and technical changes within its lifetime accurately. However, inserting the values of the following factors is mandatory to allow the software to compare the cases by referring all possible future system costs to a present value. It is noteworthy that changes in any of these values might change the results. Therefore, it is recommended to repeat the simulation with the most updated values while doing the study. The factors are:

1. **Nominal Discount Rate:** The nominal discount rate is the discount rate that includes the inflation rate. This value is usually determined by the national central bank and varies based on several social and economic factors. In Syria, the most updated discount rate values were 5% and 0.75% on December 31st, 2016, and December 31st, 2017. Therefore, 3% is selected as an average value.
2. **Inflation Rate:** The inflation rate is an economic term referring to a quantitative measure of the rate at which the cost for a specific product increases annually. HOMER PRO uses the inflation rate and nominal discount rate to calculate the actual discount rate given in the following equation:

$$i = \frac{i' - f}{1 + f},\qquad(6.14)$$

where

$i = $ the actual discount rate,
$i' = $ the nominal discount rate, and
$f = $ the inflation rate.

It is rare to have a negative interest rate. Therefore, the value of the inflation rate is limited to values less than 3%. So, 2% was the rate of choice. Within the last decade, the historical inflation rate shows that the inflation rate was relatively consistent before the Syrian war, i.e., before 2011, in the range of 1–3%. A range to which it is returned nearing the war's ends, as seen in Fig. 6.27.
3. **Annual Capacity Shortage:** The maximum allowable value of the annual capacity shortage can be found by using Eq. 6.15. This value was set to zero since the street lighting system is a critical application due to its safety role for pedestrians and drivers.

$$Annual\ Capacity\ Shortage = \frac{Total\ capacity\ shortage}{Total\ electric\ load}.\qquad(6.15)$$

4. **Project Lifetime:** A project lifetime is determined to be 25 years based on the longevity of the PV module, which has the most extended lifespan among all system components.

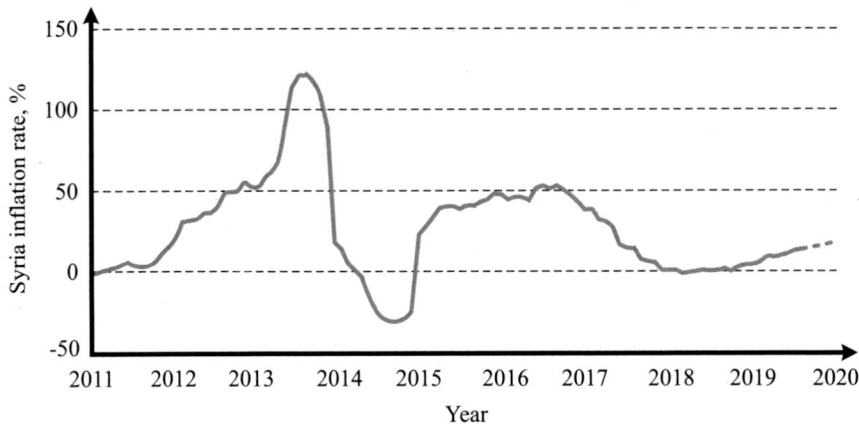

Fig. 6.27 Syrian inflation rate for the last decade [18]

Table 6.18 Project specification

Nominal discount rate (%)	3
Inflation rate (%)	2
Annual capacity shortage (%)	0
Project lifetime (years)	20

Fig. 6.28 System schematic

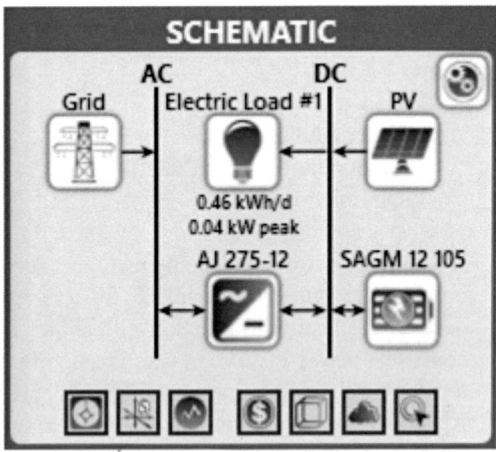

The overall values taken for the designed case are shown in Table 6.18, and the overall schematic of the proposed system is shown in Fig. 6.28.

6.4.2.8 Economic and Sensitivity Analysis of the Designed Case Studies

The results of the HOMER simulation operations show that none of the 12 cases were more financially feasible than the basic case. Thus, sensitivity and economic analysis were also conducted using HOMER Pro Software to find the critical grid rate, after which the PV and energy storage option becomes more feasible. Tables 6.19 and 6.20 illustrate the results for the M3 and M4 lighting classes, respectively. Based on the tables, the bifacial modules with 40% albedo and the lithium-ion battery in the M4 lighting class is the most economical case. The most economical case has been highlighted in Table 6.20.

6.5 Conclusion

This chapter spans three main parts: design of a standalone PV system, design of a hybrid PV system, and design of distributed PV systems. Starting with a general discussion on standalone PV systems, the chapter presents a detailed design process of an off-grid PV system based in Indonesia. Then the design process of a hypothetical hybrid PV system has been described. The use of distributed solar PV systems covers two major applications: irrigation pumps in agriculture and street lighting systems. Each application is explained in the light of some practical case studies that will help the reader to grasp the concept well. After completing this chapter, the reader will get first-hand knowledge of designing standalone, hybrid, and distributed PV systems. In the next chapter, you will find the design process of grid-tied solar PV systems, the negative aspects of such a system, and the challenges and limitations barring the flourish of such systems in developing countries.

6.6 Exercise

1. What is a hybrid PV system? What are the main components of such a system?
2. What are the benefits of standalone or off-grid PV systems?
3. Design an off-grid PV system for a school with 100 kWh daily energy consumption.
4. Design a hybrid PV system for a house with 25 kWh daily energy consumption.
5. How can solar PV systems be used in agricultural applications?
6. Discuss how solar PV street lighting systems are designed.

Table 6.19 Economic and sensitivity analysis for M3 class

Case	NPC($)	COE($)	Operating cost ($/year)	Initial capital ($)	$\frac{NPC\ of\ the\ Case}{NPC\ of\ the\ Base\ case}$ (%)	Critical grid price ($)	Grid price raise (%)
Monofacial module(s) Lead acid							
Grid	977.53	0.198	42.24	45	100	–	92.6
PV and batteries	1788	0.362	27.95	1171	182	0.316	
Bifacial module(s) (20% Albedo) Lead acid							
Grid	977.53	0.198	42.24	45	100	–	93.3
PV and batteries	1793	0.363	41.47	877.92	183	0.317	
Bifacial module(s) (40% Albedo) Lead acid							
Grid	977.53	0.198	42.24	45	100	–	91.4
PV and batteries	1777	0.359	41.41	862.67	181	0.314	
Monofacial module(s) Lithium ion							
Grid	977.53	0.198	42.24	45	100	–	41.4
PV and batteries	1340	0.271	-35.02	2113	137	0.232	
Bifacial module(s) (20% Albedo) Lithium ion							
Grid	977.53	0.198	42.24	45	100	–	42
PV & batteries	1342	0.271	-35.11	2117	137	0.233	
Bifacial module(s) (40% Albedo) Lithium ion							
Grid	977.53	0.198	42.24	45	100	–	40
PV and batteries	1331	0.269	-35.14	2107	136	0.231	

Table 6.20 Economic and sensitivity analysis for M4 class. The case in blue font is the most economical case

Case	NPC($)	COE ($)	Operating cost ($/year)	Initial capital ($)	$\frac{NPC \ of \ the \ Case}{NPC \ of \ the \ Base \ case}$ (%)	Critical grid price ($)	Grid price raise (%)
Monofacial module(s) Lead acid							
Grid	753.73	0.203	32.23	42.19	100	–	88
PV & batteries	1333	0.359	24.26	797.47	176	0.309	
Bifacial module(s) (20% Albedo) Lead acid							
Grid	753.73	0.203	32.23	42.19	100	–	89
PV and batteries	1335	0.36	24.19	800.58	177	0.31	
Bifacial module(s) (40% Albedo) Lead acid							
Grid	753.73	0.203	32.23	42.19	100	–	87
PV and batteries	1324	0.357	24.15	791.12	175.6	0.307	
Monofacial module(s) Lithium ion							
Grid	753.73	0.203	32.23	42.19	100	–	53
PV and batteries	1099	0.296	-43.98	2070	145.8	0.251	
Bifacial module(s) (20% Albedo) Lithium ion							
Grid	753.73	0.203	32.23	42.19	100	–	53
PV and batteries	1099	0.296	-44.04	2071	145.8	0.251	
Bifacial module(s) (40% Albedo) Lithium ion							
Grid	753.73	0.203	32.23	42.19	100	–	26.8
PV & batteries	928.05	0.25	-8.56	1117	123	0.208	

References

1. Characteristics of lead acid batteries (2021). https://www.pveducation.org/pvcdrom/lead-acid-batteries/characteristics-of-lead-acid-batteries
2. 5 kva 4000 w hybrid pure sine wave inverter mps-5k solar hybrid sine wave inverter with 60a 48v mppt controller (2021). https://www.amazon.com/4000W-Hybrid-Inverter-MPS-5K-Controller/dp/B082DBWPCM
3. Philadelphia Solar, Ps-p60-(275-290 w) poly crystalline module (2021). https://www.philadelphia-solar.com/uploads/P60_outline.pdf
4. Q.peak duo l-g5.2 380–395, https://www.q-cells.com/dam/jcr:1f56c107-021b-4e49-a648-c988accdc946/Hanwha_Q_CELLS_Data_sheet_QPEAK_DUO_L-G5.2_380-395_2018-03_Rev05_NA.pdf
5. Mighty max 12v 250ah, https://www.mightymaxbattery.com/shop/solar-panel-batteries/12v-250ah-sealed-lead-acid-battery-for-scada-systems-solar-backup/
6. Easun power hybrid solar inverter 3kw 220v 24v 5500w pv 450vdc input 90a mppt grid tied touch screen inverter with ct sensor. https://www.easunpower.com/products/EASUN-POWER-Hybrid-Solar-Inverter-3KW-220V-24V-5500W-PV-450Vdc-Input-90A-MPPT-Grid-Tied-Touch-Screen-Inverter-With-CT-Sensor
7. U. Sharma, S. Kumar, B. Singh, Solar array fed water pumping system using induction motor drive, in *2016 IEEE 1st International Conference on Power Electronics, Intelligent Control and Energy Systems (ICPEICES)* (IEEE, Piscataway, 2016), pp. 1–6
8. N. Muñoz-Galeano, O.A. Arraez-Cancelliere, J.M. López-Lezama, *Wind Solar Hybrid Renewable Energy System*, chapter Methodology for Sizing Hybrid Battery–Backed Power Generation Systems in Off-Grid Areas (IntechOpen, 2020)
9. S. Govindasamy, M. Yogaraj, V. Mahes Kumar, M. Bhuvanesh, Design and Implementation of Solar PV FED BLDC Motor Driven Water Pump Using MPPT 4 (2018)
10. LORENTZ website (2021). https://www.lorentz.de/
11. National Fire Protection Association (NFPA), *NFPA 70 - National Electric Code*. (2020)
12. MTE Corporation, *dV/dT Filter Series A 440–600 VAC User Manual* (2004)
13. dv/dt filter manual, MTE corporation (2021). https://www.mtecorp.com/pages_lang/wp-content/uploads/INSTR-019Rel041119dVdTFilterSeriesA440-600VACUserManual.pdf
14. Transcoil.com (2021). https://transcoil.com/support/faq/
15. Time and Date. Damascus, Syria—sunrise, sunset, and daylength (2020). https://www.timeanddate.com/sun/syria/damascus?month=12&year=2020
16. Ps-m72-(370-385w) (2021). https://www.philadelphia-solar.com/uploads/M72_outline.pdf
17. Ps-m72(bf)-(370-385w) (2021). https://www.philadelphia-solar.com/uploads/BF_datasheet_M72_M2-Size-_outline.pdf
18. Syria inflation rate (2021). https://tradingeconomics.com/syria/inflation-cpi#

Design of Grid-Tied PV Systems

7

7.1 Introduction

The conventional electricity grid in most countries of the world is still fed by electricity generated using conventional energy sources—fossil fuels. Nuclear energy has also garnered significant attention as well as controversies as a source of clean electricity. Fossil fuels have two major drawbacks—limited reserves and liberation of greenhouse gases leading to global warming and climate change. For averting climate change, achieving net-zero emissions and limiting the global temperature rise below 2 °C according to the Paris Agreement are the prime concerns of the global community. Due to the clean energy from the Sun and the ample availability of sunlight, integrating PV power into the grid is a viable option at present. Grid-tied PV systems are PV systems that are integrated into the utility grid through a suitable DC/AC conversion mechanism. Synchronization is the prerequisite for injecting the AC power derived from the DC power of the solar PV arrays. For synchronization, the voltage magnitude, frequency, phase angle, phase sequence, and the waveform of the output AC power should be matched with that of the utility grid.

Grid-tied PV systems have the major share in the solar industry market due to several advantages including, but not limited to:

- Energy storage devices, such as batteries, are not necessary for these systems. Batteries add up to the system costs and are also subject to replacement over the project lifetime (typically 25 years).
- Grid-tied systems do not have a maximum or minimum rating, and they can range from a few kW in residential applications up to a few GW in utility-scale power plants.
- Most of these systems are subsidies by tax credits or high purchase rates that shorten the payback period to the range of 5–10 years.
- There is good scope for future expansion. These systems might start small due to budget limits and grow later without losing any initial system components.

© The Author(s), under exclusive license to Springer Nature Switzerland AG 2022
Y. Abou Jieb, E. Hossain, *Photovoltaic Systems*,
https://doi.org/10.1007/978-3-030-89780-2_7

- Installing such a system might guarantee the system owner with 24/7 electricity. In some developing countries where electricity outages are part of the daily routines, the electricity supplier might sign an agreement with the installer to not cut off the electricity since the customer becomes a supplier rather than a consumer.
- The PV system owners may utilize the power for their own use before supplying to the grid. Net metering can play a vital role in this case.

7.2   Types of Grid-Tied Systems

Solar PV systems can be either on-grid, off-grid, or hybrid (grid-connected alongside a battery storage system). All these three types are employed globally to generate clean electricity. Again, grid-tied systems may be categorized into two types:

1. Grid-tied system that serves local loads and then sells to the grid: These systems have the priority to meet load requirements during the day, and any extra energy is injected into the grid. Therefore, these systems can be designed based on the load, space, and budget limits. In addition, the utility provides the system owner with a smart meter that can calculate energy flow in both directions and based on the cumulative monthly flow the customer pays or gets paid.
2. Grid-tied system that only sells to the grid: These systems are built with limited budget and spaces, whose energy yield is completely fed into the grid. Most utility-scale power plants are on-grid systems where the purchase rate is higher than the selling rate. For instance, an electricity company buys each kWh from solar by $0.11 but sells it for $0.09.

The design procedure of utility-scale solar PV systems is beyond the realm of this book. But we can certainly feast our eyes on the enormity of such a project, such as the one in Fig. 7.1.

7.3   Case Study—Sierra Cascade Nursery, Bonanza, Oregon

Sierra Cascade Nursery (SCN) [1] is located in Bonanza, Oregon, USA. In rural Oregon, it is a farm that grows strawberries, raspberries, endive root, mint root, asparagus crowns, cilantro, grain seed, and hay. In addition to growing such a wide variety of crops, the SCN is special because they have realized the importance and the benefits of a grid-tied solar PV system. The project consists of a large pole-mounted solar PV array employing a dual-axis tracking mechanism, depicted in Fig. 7.2. The project's key highlights—as provided by the Vice President of Operations at SCN, John Wells—are pointed out below.

Fig. 7.1 Grid-tied, utility-scale 2 MW solar PV farm at Oregon Institute of Technology spanning across a nine-acre area

- The grid-tied solar project is a dual-axis tracker system capable of producing 40 A, 240 V, 9.6 kW power.
- The main motivation underlying the project was to invest in something that would make a difference for the environment and have a significant return on investment (ROI).
- A one-time federal tax credit of $17,500 was provided.
- The costs for the system were $72,000 installed with an annual ROI of approximately 7–8%.
- The credit is calculated as a one-to-one credit per unit energy each month. If the energy production and consumption are equal, the bill is $0.00. However, it is not possible to produce more energy than required, as there is no battery bank. If the energy production is less, then a bill has to be paid.
- With all of the environmental and political issues regarding everything, more electricity production is required. Traditionally hydroelectric power is the best way to generate electricity and store water for several uses [2]. Unfortunately, there have been no new hydro projects for many years, so solar is another option to help with this demand.

Fig. 7.2 The 9.6 kW grid-tied solar PV project at the Sierra Cascade Nursery, Oregon, USA

Although the system is rated at 9.6 kW, it produces about 10.4 kW power in reality, thanks to its dual-axis tracking system that ensures the highest efficiency. Recall the rooftop solar PV system in Fig. 5.44. That is a 12.6 kW grid-tied system but only produces around 8–9 kW power on average. The rooftop system is much larger and less costly (only \$43,000) than the SCN system. Still, it has a lower efficiency due to the absence of a tracking mechanism and higher temperature losses. However, both the systems have a similar return on investment and payback period.

7.4 Design Steps of a Grid-Tied PV System

This section demonstrates the elaborate design process of a grid-tied system that only sells to the grid. The system under design has been implemented in Syria, as shown in Figs. 7.3 and 7.4, and is working well as of the time of writing this book. The system is located in the district named Hayy al Istiqlāl in the As-Suwayda Governorate in Syria, which is nearly 58 miles south of the capital, Damascus.

A preliminary study should be conducted before designing a grid-tied solar PV system. The study includes assessing the site's geography and climate and a selection of the appropriate solar module. Following the initial planning, the design of a grid-tied PV system involves several steps. The first step is the site evaluation, wherein the solar resources in the specified site are evaluated. Accordingly, the number and size of the PV arrays, along with their positioning and orientation,

Fig. 7.3 Image of the 100 kW grid-tied solar PV system installed in Hayy al Istiqlāl, Syria

Fig. 7.4 Image from under the 100 kW grid-tied solar PV system installed in Hayy al Istiqlāl, Syria

are determined. The inverter, cabling, protective equipment, shading analysis, grounding, net metering, storage facilities, etc., are also specified during the design stage. These steps require thorough calculations and prudent judgment to ensure that the designed system works according to plan and accrues the desired benefits. After the system design, a simulation is run to analyze the performance of the designed system. The performance analysis has to evaluate the energy production, system losses, and the performance ratio or efficiency.

7.4.1 Site Evaluation

The designer should start with the site evaluation process wherein the temperature, solar altitude, solar irradiance data acquisition process, and shading analysis are done. The tilt angle, array spacing, and mounting system are also determined to maximize the energy yield along with the minimum surface occupation.

Usually, the monthly global horizontal irradiance (GHI) data for a specific location can be obtained from the nearest weather station. Using the data, the global tilted irradiance (GTI) can be easily obtained using the MATLAB code provided previously in Sect. 4.16, or solar irradiance websites could be directly utilized to obtain the best tilt angle at which solar modules receive the highest solar irradiance. This step is mandatory to determine the distance between the arrays (pitch) using Eq. 7.1.

$$D = H \times \frac{\cos \alpha}{\tan \beta},\tag{7.1}$$

where,

D is the minimum distance between the arrays,
H is the height of the array,
α is the angle between the solar azimuth and the array azimuth, and
β is the solar altitude angle.

Shading is an unwanted factor for solar PV modules. The solar arrays should be placed maintaining a minimum distance D to avoid shading one module by its neighboring modules. Both α and β vary with the hour of the day and the month of the year. Typically, these values on 21st December are determined using a Sun path diagram. The height of the array H depends on the tilt and size of the modules, which is a function of the latitude of the location and the power requirements, respectively. The larger the module size, the larger is the value of H. Figure 7.5 shows the determination of the minimum row distance or array spacing for a PV system. Equation 7.1 can also be utilized to calculate the shadow length of some objects such as poles, trees, or buildings, so that the designer can avoid the possible shading areas.

Since solar modules are heavily dependent on temperature, the solar designer should consider the extreme high and cold temperatures for a decent amount of time. Most weather stations provide temperature data for standard 5, 10, 20, and 50

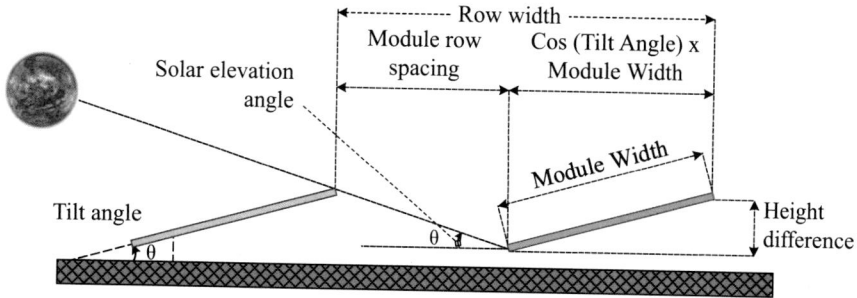

Fig. 7.5 Determination of the minimum row distance or array spacing for a PV system

Table 7.1 Temperature rise based on the mounting structure [4]. The value of T_{rise} is taken from this table

Mounting structure	Temperature rise above the surrounding, °C
Tilted roof	32
Flat roof	36
Pole mount	29
Ground mount	30

years. The designer should study at least 20 years of temperature records to give a good estimation of the operating conditions of the system.

The mounting system of choice also has a significant role in heating the modules, where the temperature rise varies based on the mounting structure [3], as illustrated in Table 7.1.

7.4.2 Solar Array Sizing

After determining the size and budget limits, the power rating of the solar modules can be easily calculated based on the available space, tilt angle, shading analysis, and mounting structure. After that, the series and parallel configuration of the solar modules should be carefully determined based on the inverter of choice that is supplied with a very detailed datasheet through which the designer can select the minimum and the maximum number of strings connected in parallel and the number of modules in each string or (PV source circuit). The sizing process of the solar modules is elaborately described in the following subsections.

7.4.2.1 Minimum Number of Modules per String

To supply the inverter with sufficient voltage that guarantees the minimum voltage requirements of the power electronics elements in the inverter, all kinds of losses should be taken into consideration, such as temperature, voltage drop, and degradation. Therefore, the minimum number of modules per string is calculated based on the V_{mpp} of the modules of choice, the highest expected operation temperature, and a safety degradation factor. The safety degradation factor is used to simulate the possible module degradation within the lifespan of the inverter (which is usually expected to be 10 years). Therefore, the minimum number of modules per string is calculated using the following equation:

$$Minimum\ string\ voltage \geq 112\% \times inverter's\ minimum\ input\ voltage. \tag{7.2}$$

Here, 12% is added to overcome the possible voltage degradation within 10 years of the inverter's lifespan. The minimum number of modules in each string can be calculated from the following equation:

$$Minimum\ number\ of\ modules\ per\ string =$$
$$= \frac{112\% \times inverter's\ minimum\ input\ voltage}{Actual\ MPP\ voltage\ of\ each\ module}. \tag{7.3}$$

The value of the denominator of Eq. 7.3 can be determined using the following equation [5]:

$$Actual\ MPP\ voltage\ of\ each\ module$$
$$= V_{mppSTC} \times [(V_{TC} \times (T_{rise} + T_{max} - 25)) + 100\%], \tag{7.4}$$

where:

V_{mppSTC} is the voltage at maximum power point measured at the standard test conditions (V),

V_{TC} is the voltage temperature coefficient (%/°C),

T_{rise} is the temperature rise from the mounting system (°C), which is taken from Table 7.1, and

T_{max} is the maximum expected temperature (°C) obtained from the ASHRAE website [6].

7.4.2.2 Maximum Number of Modules per String

In contrast with the previous case, a cold atmosphere leads to a voltage rise in the string, which should not exceed the inverter's maximum input voltage. Therefore, the lowest possible temperature is desired to avoid any overvoltage fault that will cause a shutdown in the inverter. Usually, this case occurs during the early mornings of sunny and cold days. The minimum number of modules per string is calculated using the following equation:

$$Maximum\ string\ voltage \leq Inverter's\ maximum\ input\ voltage. \tag{7.5}$$

The maximum number of modules in each string can be calculated from the following equation:

$$Maximum\ number\ of\ modules\ per\ string$$

$$= \frac{Inverter's\ maximum\ input\ voltage}{Actual\ OC\ voltage\ of\ each\ module}. \tag{7.6}$$

The value of the denominator of Eq. 7.6 can be determined using the following equation:

$$Actual\ OC\ voltage\ of\ each\ module$$

$$= V_{ocSTC} \times [(V_{TC} \times (T_{min} - 25)) + 100\%], \tag{7.7}$$

where:

V_{ocSTC} is the open-circuit voltage at the standard test conditions (V),
V_{TC} is the voltage temperature coefficient (%/°C), and
T_{min} is the minimum expected temperature (°C) obtained from the ASHRAE website [6].

7.4.2.3 Number of Strings in Parallel

After determining the maximum and the minimum number of modules per string, a number of strings in parallel can be calculated by checking the rated range of the input DC current that is always supplied by the inverter manufacturer.

$$Number\ of\ strings = \frac{Inverter's\ rated\ input\ current}{Module's\ I_{mpp}}. \tag{7.8}$$

7.4.3   Inverter Selection and DC/AC Ratio

In addition to the pure sine wave output signal that the inverter should have, it also has to be compatible with the grid in terms of the frequency, voltage, and phase for grid synchronization. Within microseconds after grid synchronization, the inverter output voltage should be higher than the grid voltage so that power can flow from the inverter to the grid. A slight difference in voltage is a prerequisite for the flow of power.

Due to all the system losses (temperature, dust, voltage drop, and mismatch), a lot of studies have been conducted to determine the best *DC/AC ratio*, which is the ratio of the nominal power rating of the PV module to that of the inverter. This ratio varies between 1 to 1.3 based on economic studies conducted to maximize the levelized cost of energy (LCOE) and minimize the system's initial cost.

When the DC/AC ratio of the system is less than 1, all the generated energy by the modules is supplied to the grid. However, when the ratio is greater than 1, some peak energy is clipped by the inverter and considered as losses. The optimum

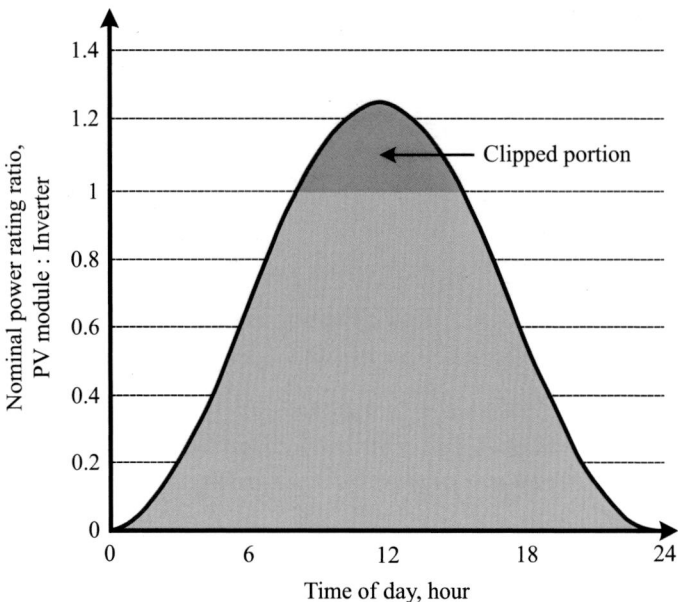

Fig. 7.6 The clipping effect when the DC/AC ratio is greater than 1

DC/AC ratio varies based on factors such as the cost of the inverters and modules. Figure 7.6 demonstrates the clipped portion when the DC/AC ratio exceeds 1.

7.5   Techno-Economic Analysis of Grid-Tied System

To explain how a grid-tied system is designed, this section describes an example of the design of a 100 kW grid-tied system using 330 W polycrystalline modules and two 50 kW string inverters. The electrical specification and dimensions of the inverter and the PV modules can be found in Tables 7.2 and 7.3. These specifications are to be utilized in the sizing process.

In the inverter datasheet, the AC grid connection type 3W + N + PE implies a 3-phase AC 5-wire system, with 3 live wires (W), one neutral (N) conductor, and one protective earthing (PE) conductor.

7.5.1   Site Evaluation

The site for designing the 100 kW grid-tied system in Syria has a 240 m^2 house where its roof could be utilized to install 102 modules. The remaining 202 modules are installed on a ground mounting structure in the backyard of the house. Both subarrays are mounted on a ground mounting structure with a tilt angle of 30°

Table 7.2 Specification of the MAC 50KTL3 LV inverter by Growatt [7]

Parameter	Value
DC side	
Maximum DC power (W)	65,000
Maximum DC voltage (V)	1100
Start voltage (V)	250
Nominal voltage (V)	600
MPPT voltage range (V)	200–1000
Maximum input current per MPPT (A)	50/37.5/37.5
Number of MPPT/string per MPPT	$3/4 + 3 + 3$
AC side	
Rated AC output power (W)	50,000
Maximum AC apparent power (VA)	55,500
AC Nominal voltage (V)	230/400
AC grid frequency (HZ)	50/60
Maximum output current (A)	80.5
Power factor	0.8 Leading and lagging
THD	$<3\%$
AC grid connection type	$3W + N + PE$
Maximum efficiency	98.7%

Table 7.3 Specification of the PS-P72-(330-345w) Polycrystalline PV module by Philadelphia Solar [8]

Parameter	Value
Open-circuit voltage, V_{oc} (V)	45.75
Short-circuit current, I_{sc} (A)	9.19
Maximum power voltage, V_{mpp} (V)	37.52
Maximum power current, I_{mpp} (A)	8.8
Maximum power, P_{max} (W)	330
Module efficiency, %	16.9
Voltage temperature coefficient, (%/°C)	−0.32
Current temperature coefficient, (%/°C)	+0.05
Power temperature coefficient, (%/°C)	−0.4
Annual degradation	−0.7%
Dimensions (mm)	$1968 \times 990 \times 40$

since the location has the latitude of 30° at which the GTI is maximized. The house orientation is already perfect for solar installation since it is faced to the south.

The Sun path of the mentioned location is extremely important in this study. The designer should calculate the minimum spacing between the rooftop array and the one in the backyard to avoid shading in particular hours in December, which will minimize the shading losses during the rest of the year. Figure 7.7 shows the Sun path diagram of Hayy al Istiqlāl, Syria. The modules will be partially shaded anytime below the horizontal dashed line. For instance, the modules will be partially shaded on June 22 from sunrise to 6:45 a.m. and from 4:45 till sunset.

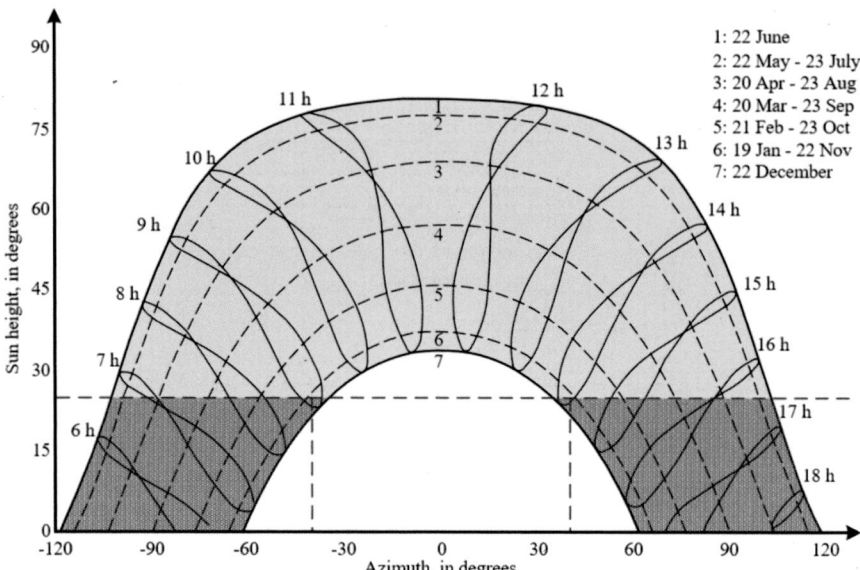

Fig. 7.7 The Sun path diagram in Hayy al Istiqlāl (32.7058° North, 36.5652° East), Syria

The 102 modules on the rooftop are designed to be in 6 rows × 17 columns, which will expand over an area of 198.7 m². The height of the array is 10.22 m, and that should be added to the height of the house, which is 4 m. Therefore, the total height that should be inserted into Eq. 7.1 is 14 m. The window from 9 a.m. to 2 p.m (5 h) on December 22nd (the shortest day of the year and the day on which the Sun is at the lowest altitude from the horizon) is supposed to be free of shading losses to the below array. Thus, the solar altitude is selected to be 25°, and the azimuth is 40°. As a result, the minimum distance between the arrays is 23 m, as Eq. 7.9 illustrates.

$$D = H \times \frac{\cos \alpha}{\tan \beta} = 14 \times \frac{\cos 40}{\tan 25} = 22.9 \; m \approx 23 \; m. \tag{7.9}$$

7.5.2 Temperature Information

In order to select the PV modules and inverters configurations, the minimum and maximum temperature values for the last 50 years should be taken into consideration. Therefore, the ASHRAE website [6] is utilized to obtain this information from the nearest weather station. This website shows that the minimum and maximum temperatures at the said location are −11.8 °C and 46.3 °C, respectively. The temperature forecast for the next 25 years is also recommended to consider since the global temperatures are increasing significantly at present.

7.5.3   PV Modules and Inverter Distribution

First, let us determine the minimum number of modules per string after which the
string can operate the inverter. The minimum input voltage of the inverter could be
found from Table 7.2 as 250 V. Using Eq. 7.4, we get:

$Actual\ MPP\ voltage\ of\ each\ module$

$$= V_{mppSTC} \times [(V_{TC} \times (T_{rise} + T_{max} - 25)) + 100\%],$$

$$= 37.52 \times [(-0.32\% \times (30 + 46.3 - 25)) + 100\%] = 31.3\ V. \qquad (7.10)$$

Next, we need to find the maximum number of modules per string to determine
the other limit.

$$Minimum\ number\ of\ modules\ per\ string$$

$$= \frac{112\% \times inverter's\ minimum\ input\ voltage}{Actual\ MPP\ voltage\ of\ each\ module}$$

$$= \frac{112\% \times 250}{31.3} = 8.9\ modules = 9\ modules. \qquad (7.11)$$

$Actual\ OC\ voltage\ of\ each\ module$

$$= V_{ocSTC} \times [(V_{TC} \times (T_{min} - 25)) + 100\%]$$

$$= 45.75 \times [(-0.32\% \times (-11.8 - 25)) + 100\%] = 51.1\ V. \qquad (7.12)$$

$$Maximum\ number\ of\ modules\ per\ string$$

$$= \frac{Inverter's\ maximum\ input\ voltage}{Actual\ OC\ voltage\ of\ each\ module}$$

$$= \frac{1100}{51.1} = 21.5\ modules = 21\ modules. \qquad (7.13)$$

To facilitate the installation process for the installers, each string will have 16
modules, and the system will be composed of 19 strings; the designer has the
freedom to pick any number between 9 and 21 to compose a string, so the number
of strings can be different without any problems. In this particular case, where the
inverter is a string inverter that has multiple MPPT, and each MPPT has its own
inputs, it is possible to have unmatched strings, each having a different number of
modules. However, we will not have this approach in this case; both strings will
have the same number of modules in this design.

Here, both the inverters have 10 inputs each. For inverter 1, each of the inputs is
connected to a string having 16 modules. For inverter 2, nine inputs are connected

with a string having 16 modules. This is the overall structure of the PV modules and their connection to the two inverters.

7.5.4 Overcurrent Protection and Circuit Breakers

In order to protect the strings from any short circuit and provide the appropriate disconnect switch that facilitates the possible maintenance process, fuses and circuit breakers (CB) should be installed before the connection point of the inverter input and the string. Each of the mentioned elements should hold 156% of the short-circuit current of the string, which has a value of 9.19 A. The minimum accepted rating of these devices is ($9.19 \times 1.56 = 14.3364$ A). So, the nearest overcurrent protection device and CB standard rating is 15 A, which is the chosen size.

For the AC side, the inverter has built-in AC protection. Thus, the designer should just do the system sizing of a three-phase CB on the output side of the inverter. This CB should handle up to 125% of the maximum output current that could be supplied by the inverter, which is provided in the datasheet as 80.5A. The minimum accepted CB rating is ($80.5 \times 125\% = 100.625$ A). In such a case, both 100 A or 110 A are accepted based on the market availability.

The output of the inverters is usually combined using switchgear or a Copper busbar based on the national standard of the installation location regulations. The output of the switchgear could also contain fuses and circuit breakers with a minimum current rating of the overall output current of the two inverters with a 1.25 times safety margin to not overload it by over 80% of its rated capacity. After that, the AC output is connected to two meters in series to avoid any problems with the meter's calibration. The utility usually paid the installer based on the higher reading if they are not equal.

7.5.5 Cable Sizing

The cable needs to be sized for two places—from the PV module to the inverter and from the inverter to the distribution board. For each cabling, the maximum ampacity and the allowable voltage drop need to be considered. For cable sizing, the NEC 2020 Handbook will be exclusively used, which is provided in Sect. 5.6.1.3.

7.5.5.1 PV Module to Inverter
Each string has a short-circuit current of 9.19 A, and the cable ampacity should hold 156% of the string short circuit. Therefore, the current at which the cable should be sized is $9.19 \times 1.56 = 14.3364$ A. Therefore, according to Table 310.16 of NEC 2020, at 75 °C, a copper THHW cable of size AWG #14 is sufficient since it handles up to 20 A. However, the ambient temperature might reach 46–50 °C. Therefore, this cable can handle 0.75×20 A $= 15$ A current, as per Table 310.15(B)(1) of NEC 2020. So, the AWG #14 wire size is adequate for this case.

The voltage drop should also be taken into consideration where the nominal voltage of each string ($16 \times V_{mpp} = 16 \times 37.52\ V = 600.32\ V$). The voltage drop on the cable that has the maximum length of 40 m (=131.234 ft) should not exceed $2\% \times 600.32 \approx 12$ V. Since the distance is 131 ft, the two-way wiring will involve a wire of $131 \times 2 = 262$ ft $= 0.262$ kft. For AWG #14 wire, the resistance of the wire is 3.14 Ω/kft, according to Table 8, Chapter 9 of NEC 2020. So, the voltage drop within the AWG #14 wire size will be:

$$V_{drop} = I \times R_C \times L = (1.25 \times 8.8) \times 3.14 \times 0.262 = 9.049\ V,$$

which is less than the permitted voltage drop of 12 V. Thus, the wire size AWG #14 copper can be deemed correct.

7.5.5.2 Inverter to AC Distribution Panel

Since $I_{max} = 80.5$ A, the cable on the output side of the inverter should have a minimum ampacity of ($80.5 \times 1.25 = 100.625$ A). The temperature effect was not taken into consideration since the inverter is installed indoors in a relatively cool area where the temperature will not exceed 30 °C. Thus, from Table 310.16 of NEC 2020, at 75 °C temperature, the copper THHW AWG #2 wire that holds up to 115 A is sufficient.

It is assumed that the voltage drop of the inverter [7] should not exceed 1.5% between the inverter output and the meters and distribution panel, which has a distance of 4 m $= 13.1234 \approx 13$ ft $= 0.13$ kft. Since the inverter voltage is 400 V_{AC}, so the limit is $400 \times 0.015 = 6$ V. For AWG #2 wire, the resistance of the wire is 0.19 Ω/kft, according to Table 9, Chapter 9 of NEC 2020. So, the voltage drop can be calculated as:

$$V_{drop} = 100.625 \times 0.19 \times 0.13 \times 2 = 4.97\ V,$$

which is less than 6 V. So, the cable size of AWG #2 is correct for this case.

7.5.6   Grounding

The rating of the overcurrent protection device chosen in this design is 15 A. According to Table 250.122 of NEC 2020 [9], the minimum size of the equipment grounding conductor (EGC) is AWG #14 copper wire based on the overcurrent protection device of choice.

7.6   Project Simulation Using PVsyst

In the previous section, we were able to obtain the following:

- General arrangement of the system.

- Pitch distance (distance between arrays).
- Lowest and highest possible temperature on site.
- String size and number of strings in parallel.
- AC and DC cable sizes based on the ampacity and voltage drop limitations.
- Overcurrent protection devices in addition to the needed grounding cable.

With that being determined, we have the needed information to run a techno-economic study using an available and reliable software that is widely used in the industry, to understand the financial and technical feasibility of any project. The outcomes of such a study will be the main reference using which the owner or developer of the project will decide on proceeding with the EPC (Engineering, Procurement, and Construction) operations.

In this section, the PVsyst software is utilized to simulate the technical and financial behavior of the system, which includes all possible losses that occurred due to the modules' aging or defects, ohmic losses, thermal losses, light-induced degradation (LID), shading objects, inverter clipping if the DC/AC ratio is above 100%, inverter auxiliary loads, etc.

7.6.1   PVsyst Output Charts, Values, and Tables

Table 7.4 illustrates the estimated energy yield at the direct output of the arrays and at the point of interconnection (POI) where the AC ohmic losses, the possible unused energy due to high DC/AC ratio where the inverter clips the energy that exceeds its power rating, inverter efficiency, and auxiliary losses are all taken into consideration.

Table 7.4 Monthly energy yield at the array side and at the POI of the 100 kW grid-tied PV system

Month	Energy at the output of the arrays, MWh	Energy at the output of the inverter, MWh
January	10.61	10.45
February	9.38	9.24
March	13.13	12.95
April	14.94	14.74
May	18.81	18.57
June	18.69	18.44
July	19.89	19.63
August	20.08	19.84
September	18.65	18.42
October	17.35	17.14
November	14.89	14.70
December	11.93	11.77
Annual total	188.35	185.89

Table 7.5 Energy yield within the project lifespan of the 100 kW grid-tied PV system

Year	Energy yield at array, MWh	Energy yield at POI, MWh
0	188	186
1	187	185
2	186	183
3	184	182
4	183	181
5	182	179
6	180	178
7	179	177
8	178	175
9	176	174
10	175	173
11	174	172
12	173	170
13	171	169
14	170	168
15	169	166
16	167	165
17	166	164
18	165	162
19	163	161
Total	3516	3471

Based on the estimated energy yield and the annual degradation rate (-0.7%) that the solar modules' manufacturer provides in the datasheet, we can obtain the annual energy yield at the end of each year within the project lifetime, which is basically determined by the duration of the power purchase agreement (PPA) with the balancing authority or the utility grid operator. This duration is usually set to 20 or 25 years. In this case, the project lifetime is 20 years. For each year, the energy output from the PV array and that at the POI is enlisted in Table 7.5.

Figure 7.8 illustrates the gain/loss flow chart (from the PVsyst software) through which the reader can check the losses and possible gains in each phase of the energy conversion process starting from the solar irradiance and ending up when AC power is being injected into the point of interconnection. Through this chart, the designer can obtain the Net Capacity Factor (NCF), through which the project owners can compare the projects' electrical components and geographical sites characteristics to determine which one is more efficient. The NCF can be defined as the ratio of the actual electrical energy output over a definite duration of time to the highest possible output of electrical energy over that period. The NCF can be calculated using Eq. 7.14, where $24 \times 365 = 8760$ is the number of hours in a year.

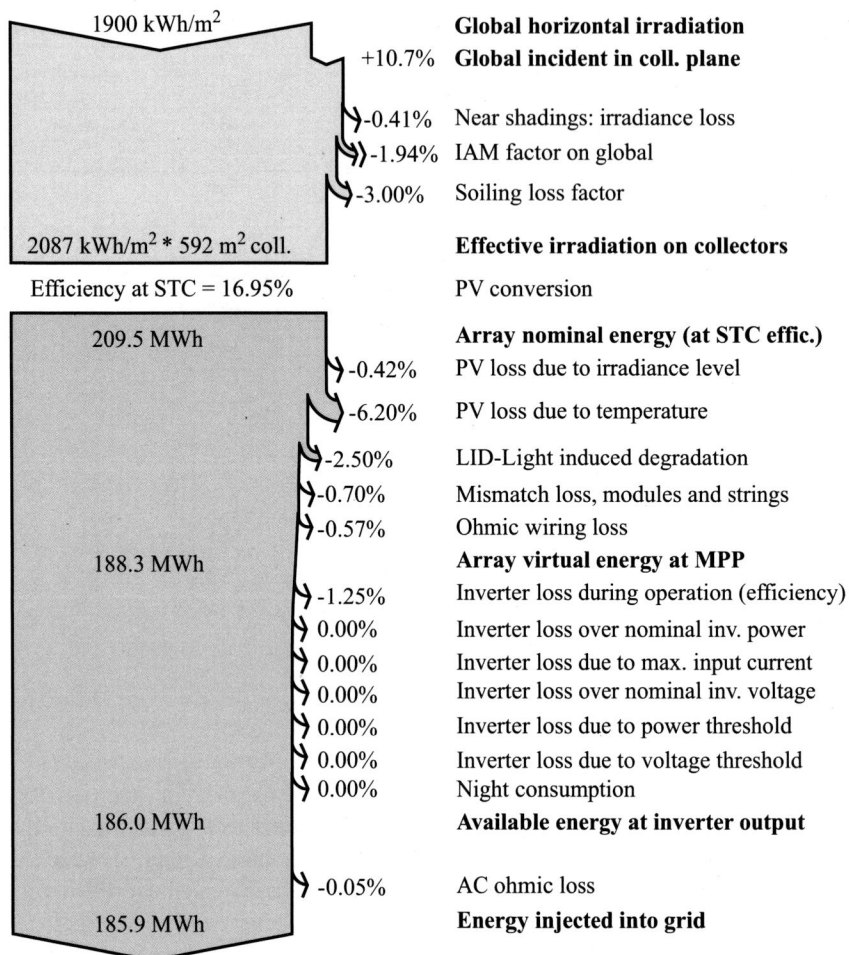

Fig. 7.8 System gain/loss diagram of the 100 kW grid-tied PV system

$$NCF = \frac{Energy\ injected\ to\ grid}{8760 \times the\ lowest\ value(AC\ capacity\ of\ inverters\ and\ grid\ limitation)}.$$
                                                                                                      (7.14)

In this case, since there is no grid limitation, the value will be 100 kW.

$$NCF = \frac{185.9 \times 1000}{8760 \times 100} = 21.22\%.$$ (7.15)

The capital, running, and replacement costs of the project are illustrated in Table 7.6. In parallel, the discount and inflation rates are set to 2% and 3%, respectively.

Table 7.6 Cost summary of the 100 kW grid-tied PV system project

Cost	Type	Amount
PV modules	Initial	$40,000
Inverters	Initial	$15,000
Wiring	Initial	$2500
Mounting system	Initial	$10,000
Surge arrestor	Initial	$500
Installation and study	Initial	$10,000
Grid connection fees and equipment	Initial	$2000
Cleaning and possible service fees	Running/O&M	$1000/year
Inverters	Replacement	$10,000 in the year 10

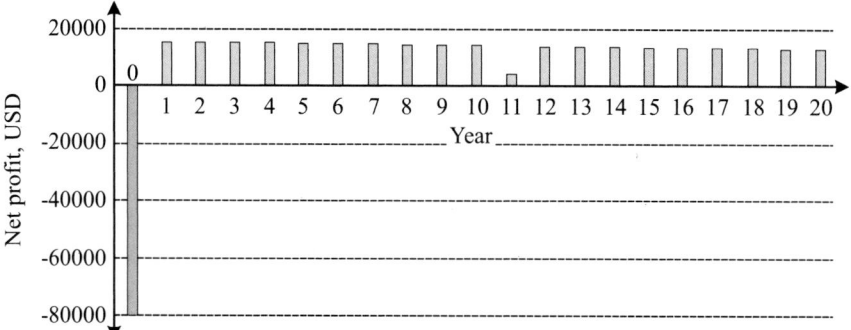

Fig. 7.9 Annual cash flow of the 100 kW grid-tied PV system

The total initial cost of the project is $80,000, and the utility grid buys each kWh by $0.11. The system is considered a low-maintenance enterprise as the O&M costs are reasonably low. The two inverters need to be replaced in year 10 with an estimated cost of $10,000. With all this information given, Table 7.7 and graphs in Figs. 7.9 and 7.10 can illustrate the breakeven point of the system.

The breakeven point of such a project with the given prices can be as short as 4.2 years. Figures 7.9 and 7.10 illustrate the annual and cumulative cash flow.

Figure 7.11 illustrates real data from this project where the cloud distortion can be seen from 8:40 a.m. till noon. However, in the afternoon period, the sky was clear, which reflects a smooth curve. This curve represents the daily energy yield of a random day in April.

7.7   Challenges of Grid-Tied PV Systems and Probable Solutions

The complete success of grid-tied PV systems is hindered by some challenges, especially during installation and integration. When PV systems are integrated

Table 7.7 Project financial summary of the 100 kW grid-tied PV system

Year	Energy yield at array, MWh	Energy yield at POI, MWh	Gross Income, USD	Running and replacement costs, USD	Net profit, USD	Cumulative profit, USD	% amount
0	–	–	–	–	–	−80,000	–
1	188	186	20,448	1000	19,448	−60,552	24.31
2	187	185	20,305	1020	19,285	−41,267	48.42
3	186	183	20,162	1040	19,122	−22,146	72.32
4	184	182	20,018	1061	18,957	−3188	96.01
5	183	181	19,875	1082	18,793	15,605	119.51
6	182	179	19,732	1104	18,628	34,233	142.79
7	180	178	19,589	1126	18,463	52,696	165.87
8	179	177	19,446	1149	18,297	70,993	188.74
9	178	175	19,303	1172	18,131	89,124	211.41
10	176	174	19,160	1195	17,965	107,089	233.86
11	175	173	19,017	11,219	7,798	114,886	243.61
12	174	172	18,873	1243	17,630	132,517	265.65
13	173	170	18,730	1288	17,442	149,959	287.45
14	171	169	18,587	1294	17,293	167,252	309.07
15	170	168	18,444	1319	17,125	184,377	330.47
16	169	166	18,301	1346	16,955	201,332	351.67
17	167	165	18,158	1373	16,785	218,117	372.65
18	166	164	18,015	1400	16,615	234,731	393.41
19	165	162	17,871	1428	16,443	251,175	413.97
20	163	161	17,728	1457	16,271	267,446	434.31
Total	3516	3471	381,762	34,316	347,446	267,446	434.31

into the utility grid, several technical problems arise, such as power quality issues, injection of harmonics, voltage, frequency, and power fluctuation, islanding, protection limitations, and so on. In addition, most on-grid solar PV systems are devoid of a storage system since the solar power produced is directly fed to the grid. This dependence on the grid can often become a problem if the grid fails due to a power shortage or other problems.

The fluctuating nature of PV output power contains a lot of harmonics. When integrated into the utility grid, the harmonics distort the shape of the AC waveform in the grid and degrade the power quality. The variable PV output may also cause voltage and frequency fluctuations in the grid power unless a reliable inverter is used that can produce an output with the specified voltage and frequency at all times.

Typically, the daily solar power output follows a rough bell-curve-like pattern and peaks at noon. For a grid-tied to PV systems, this pattern evokes a problem known as the duck curve. When PV output peaks, the demand for the grid power falls, requiring the grid to ramp down. Again, when the PV output fades away, the demand for the grid power shoots, requiring the grid to ramp up. Besides the usual

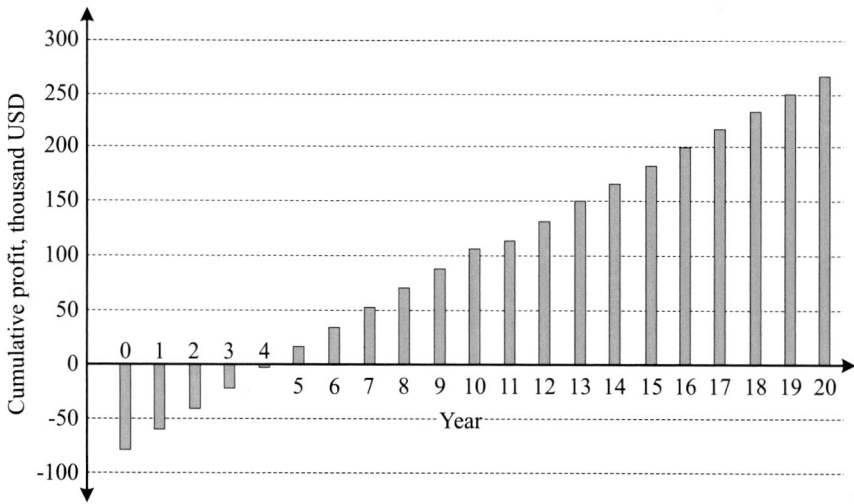

Fig. 7.10 Cumulative cash flow of the 100 kW grid-tied PV system

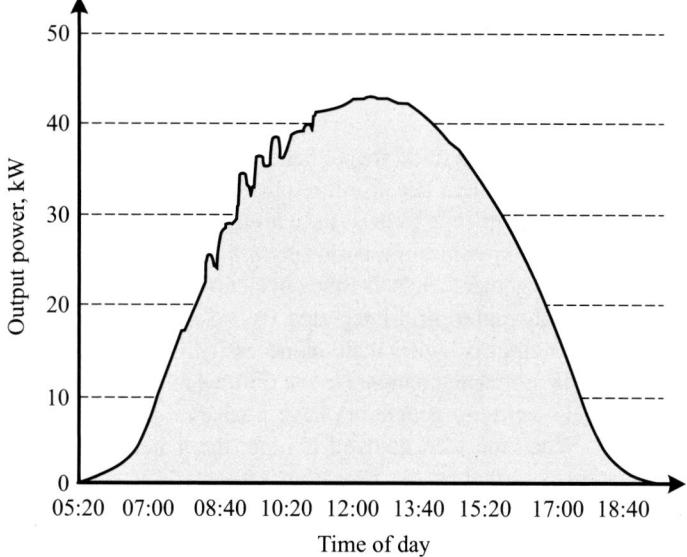

Fig. 7.11 Inverter-1 output on a random day in April

daily load curve in the absence of PV systems, the new load curve forms a duck-shaped image. This is known as the duck curve, as depicted in Fig. 7.12. Here, the orange curve is the actual load curve with an evening peak. The solar PV output is available for certain hours of the day. When solar PV output is available, the power drawn from the grid reduces, as shown by the blue curve. The orange curve and the

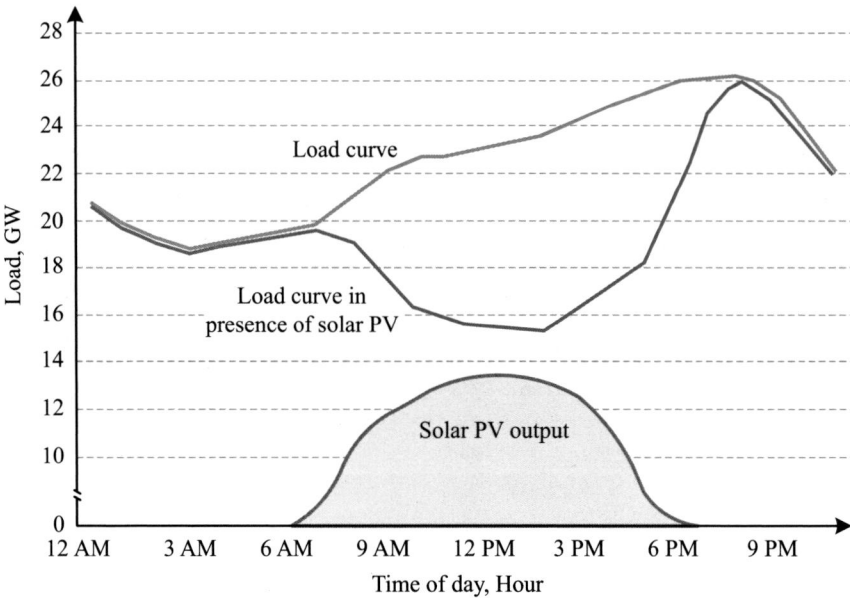

Fig. 7.12 Illustration of the duck curve phenomenon—a challenge that solar PV poses to the utility grid

blue curve together resemble a duck shape, hence the name duck curve. The duck curve phenomenon is a problem because it requires electricity-generating systems to ramp up over a very short time to keep up with the sudden rise in load demand. The use of energy storage systems, microgrids, hybridization of sources, time-of-use rates, and demand management can solve the duck curve problem.

Another problem inherent to grid-integrated PV systems is the risk of overgeneration when the PV output is higher than the necessity. Overgeneration can cause overloading and bring permanent damage to the grid, which is costly to repair.

The conventional electricity generators have a reserve capacity in the form of spinning capacity. When the load demand is high, the generator can increase its output. This feature is absent in the inverter-based solar PV systems, which are inertia-less systems having no reserve capacity unless storage devices are used. Therefore, if solar PV is integrated into the grid, a storage system is an essential component for the reliable operation of the grid.

The previous economic analysis does not consider the energy shortage that might frequently occur in developing countries, causing a relatively long payback period than expected. In Fig. 7.13, observe how the energy yield drops to zero from 11:00 a.m. to 12:00 p.m. The inverter stopped sensing the voltage signal from the grid, which leads to an immediate automatic shutdown by the inverter to stop injecting any power into the grid to avoid islanding. This protection method is mandated by all utilities to protect their linemen when doing maintenance

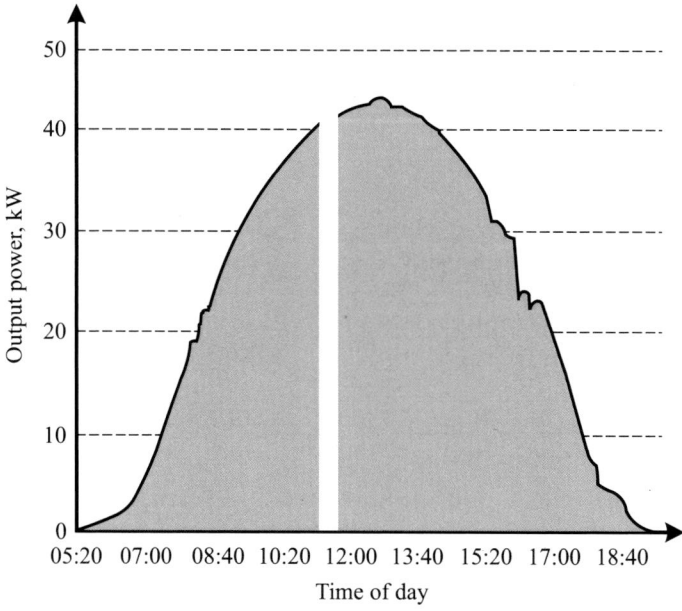

Fig. 7.13 The effect of an outage of 1 h

Table 7.8 Daily grid outage schedule in different months

Month	Daily outage time	Duration
January	9:00–11:00	2 h
February	10:00–12:00	2 h
March	11:00–13:00	2 h
April	11:00–12:00	1 h
May	13:00–14:00	1 h
June	13:00–15:00	2 h
July	13:00–15:00	2 h
August	13:00–15:00	2 h
September	13:00–14:00	1 h
October	13:00–14:00	1 h
November	11:00–13:00	2 h
December	10:00–12:00	2 h

on transmission lines and transformers where the energy from the inverters is considered as a potential source of electrical shock.

The outage schedule was simulated using HOMER Pro software that contains a powerful tool through which a designer has the freedom to determine the hour of outages during weekdays and weekends within a full year. The schedule of the outage can be seen in Table 7.8.

Table 7.9 Project financial summary with the scheduled outages

Year	Energy yield at POI with outages, MWh	Gross Income, USD	Running and replacement costs, USD	Net profit, USD	Cumulative profit, USD	% amount
0	–	−80,000	–	–	−80,000	–
1	149	16,429	1000	15,429	−64,571	19.29
2	148	16,314	1020	15,294	−49,276	38.40
3	147	16,199	1040	15,159	−34,117	57.35
4	146	16,084	1061	15,023	−19,094	76.13
5	145	15,969	1082	14,887	−4206	94.74
6	144	15,854	1104	14,750	10,544	113.18
7	143	15,739	1126	14,613	25,158	131.45
8	142	15,624	1149	14,475	39,633	149.54
9	141	15,509	1172	14,337	53,970	167.46
10	140	15,394	1195	14,199	68,170	185.21
11	139	15,279	11,219	4,060	72,230	190.29
12	138	15,164	1243	13,921	86,151	207.69
13	137	15,049	1288	13,761	99,912	224.89
14	136	14,934	1294	13,640	113,553	241.94
15	135	14,819	1319	13,500	127,053	258.82
16	134	14,704	1346	13,358	140,411	275.51
17	133	14,589	1373	13,216	153,628	292.03
18	132	14,474	1400	13,074	166,702	308.38
19	131	14,359	1428	12,931	179,633	324.54
20	129	14,244	1457	12,787	192,421	340.53
Total	2789	306,737	34,316	272,421	192,421	340.53

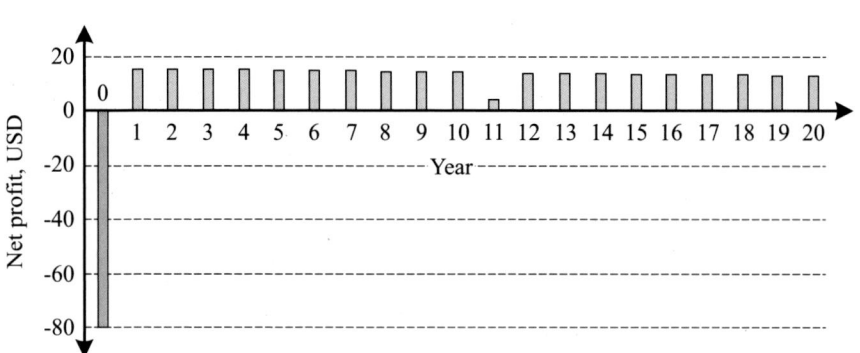

Fig. 7.14 Annual cash flow with scheduled outages

The economic analysis of the same system with the mentioned outages schedule can be summarized in Table 7.9. Moreover, the annual and cumulative cash flow can also be seen in Figs. 7.14 and 7.15, respectively.

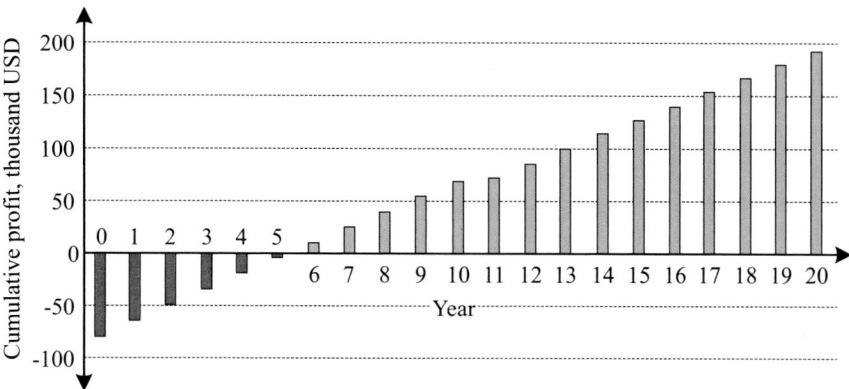

Fig. 7.15 Cumulative cash flow with scheduled outages

The breakeven point of this project with the scheduled outage is 5.4 years compared to 4.2 years in the case of the reliable grid. Therefore, on-grid PV systems may increase the payback period of the solar PV installation.

7.8   Grid-Tied PV Systems in Developing Countries: A Potential

Although grid-tied PV systems are an emerging technology in many developed nations, and a portion of the utility grid electricity is supplied by solar PV modules, this situation is not the reality in the developing countries who are slightly lagging in terms of financial and technological advancements. In this section, the solar prospects of two developing countries are discussed. Bangladesh and Syria are both countries in the northern hemisphere and receive a sufficient amount of sunlight every day. Yet, there are several challenges and limitations of the growth of solar PV technology in these countries.

7.8.1   Bangladesh

Bangladesh is a tropical country situated at 23.6850° North latitude and 90.3563° East longitude. The tropic of cancer passes across the heart of the country, which implies that the zone is subject to direct radiation for a significant time of the year. A small land having six seasons, Bangladesh enjoys ample sunshine throughout the year, and so the solar potential of Bangladesh is very high. The country has an area of 148,460 sq. km. with a hefty population of 166 million. The country is bordered by India on all its sides except at the southeast, which is shared with Myanmar, and the south is open to the Bay of Bengal.

7.8.1.1 Current Scenario of Solar PV

As of June 2021, there are 7 solar parks operational in Bangladesh with a combined capacity of 129.53 MW, 8 other solar parks under development rated 522 MW in total, and 15 other solar parks rated 925.77 MW in total are in the planning phase [10]. Solar home systems (SHS) are very popular and common in Bangladesh, with 5.8 million installations in rural areas [11]. They can be spotted in many residential and commercial buildings, both in rural and urban areas. Many homes have an SHS with battery energy storage systems as a backup power source since load shedding is a frequent occurrence in the country. The solar SHS capacity in the country is 262.55 MW. Solar street lighting systems are also implemented in several highways of the country, and there is about 17.06 MW of solar street lighting systems installed in the country. Moreover, the US developer Eleris Energy has proposed to install 2.2 GW of solar PV farms in various locations of the country, which only proves the suitability of Bangladesh as a solar power hub [12]. Solar water heating, solar irrigation pumps, solar-powered drinking water system, telecommunication, etc., are also some popular solar PV applications in Bangladesh.

7.8.1.2 Solar PV Potential

Despite having a large solar window, grid-tied PV systems have not yet flourished in this small country of more than 166 million people due to several limitations. Some measures to utilize the full solar potential in Bangladesh are as follows:

1. **Role of government initiatives:** The government of Bangladesh fell behind to take initiatives to boost the growth of grid-tied renewable energy. Incentives, subsidies, and mandates could act as a motivator for people to invest in solar PV systems.
2. **Poverty eradication:** Since most of the people of Bangladesh are below the poverty line, they cannot all afford to install SHS at their homes rated enough to feed to the grid after meeting the domestic loads. So, poverty elimination should be a priority.
3. **Decommissioning of coal and natural gas-fired power plants:** The existing power plants in Bangladesh are mostly powered by coal or natural gas, and the country already has surplus power after meeting the connected loads. So, the existing fossil fuel-based plants should be gradually decommissioned to increase the scope for solar power to grow.
4. **Rooftop solar PV systems:** Bangladesh has a small area that is insufficient to cater to the housing and farming requirements of its growing population. Hence, the land options for solar farms are quite limited. However, rooftop solar PV systems are a feasible option to boost the penetration of solar PV in the country.

7.8.2   Syria

Syria is a middle-eastern country located at 34.8021° North latitude and 38.9968° East longitude. The country has an area of 185,180 sq. km, with a population of 17 million. Syria shares its borders with Turkey on its north, Iraq on its east and southeast, Jordan on its south, and Lebanon on its southwest. The Mediterranean Sea is on the west of Syria. Syria has two main seasons—the hot and dry season from May to October and the cool and wet season from November to April.

7.8.2.1 Current Scenario of Solar PV

Before the onset of the civil war in 2011, Syria had 95% of homes connected to the utility grid. Through its network of 15 power plants, there was adequate power to feed the local loads, as well as export power to neighboring countries. Since the war broke out, infrastructure worth \$5 billion got vandalized [13]. In 2013, the Union of Medical Care and Relief Organizations (UOSSM) suggested using solar modules for powering the hospitals in Syria, as hospitals became a major necessity for the war-torn nation. At the end of 2018, there was only 1 MW of solar power installed in the country [14]. The Ministry of Electricity Transmission Establishment in Syria has planned to build two new solar plants soon—a 23 MW project in Damascus and a 40 MW project near Homs [14].

7.8.2.2 Solar PV Potential

The number one challenge for Syria in all aspects of its development is the ongoing civil war since 2011. The decade-long war has caused the loss of innumerable lives and destroyed property worth several billion dollars. It is also a decimating factor against the growth and development of the nation in terms of infrastructure, economy, education, and overall stand among the other developing nations of the world. Some positive factors that can help the growth of solar PV systems are as follows:

1. **Government initiatives:** The government of Syria is eyeing the growth of renewable energy and solar PV systems in particular. Many new projects are underway to provide power to the country, despite the decimated grid infrastructure.
2. **Infrastructure:** Years of demolition have devastated the electricity infrastructure in Syria, recovering from which is a challenge yet to be overcome. Destroyed property is a massive setback for the utility grid system. But this is a boon for the solar PV systems, as they can be installed anywhere with good solar insolation.
3. **Solar PV as an alternative to grid electricity:** The embattled country is set in miserable poverty and unemployment. Where people are struggling to manage a fair meal and a decent living, they cannot afford to pay for grid electricity. Hence, solar PV is a good option for them. The portability of solar modules also helps people to shift homes amidst the uncertainty of life.

7.8.3   Comparison with Solar PV in the USA

The USA is a developed country and ranks among the forerunners of renewable technologies, particularly solar PV technology. As of 2016, there were a million solar installations throughout the USA, which quickly doubled by 2019. In 2021, the cumulative installed capacity of solar PV in the USA is 102.8 GW_{DC}, which can power up 18.6 million homes and contributes over 3% of the total US electricity generation. Despite such a small proportion of solar energy in the energy mix, the USA is the second-largest producer of solar power, following China. California, Texas, and North Carolina are the top three US states with the highest installed capacity of solar PV systems. The other states are also going ahead rapidly in their solar PV installations.

The rapidly falling costs of solar modules, the large numbers of jobs created by the solar PV industry, the growing economy, ample availability of natural, technical, and infrastructural resources, the increasing state and federal incentives, policies, and mandates, and the overall rise in awareness about the carbon footprint of the power and energy sector, etc. have made the acceleration of the solar PV possible in the USA.

For example, the federal incentive named the solar investment tax credit (ITC) has been directly linked to a 60% rise in solar additions per annum since 2005, and the ITC currently provides 26% federal tax credit on the cost of solar installations for residential and commercial purposes.

Another popular federal policy is the SunShot Initiative, launched by the US Department of Energy (DoE) in 2011, with the purpose of making solar PV systems more affordable to the American masses. This initiative has influenced a 75% reduction in the price of solar energy. As a result, the LCOE of utility-scale solar power in 2020 stands at only $0.06/kWh, while residential and commercial costs are $0.10/kWh and $0.08/kWh, respectively. By the year 2030, this initiative targets to further lower the costs to $0.03/kWh, $0.04/kWh, and $0.05/kWh for utility, commercial, and residential applications, respectively.

In addition to the federal policies, the huge investments by the US government are also playing a powerful role in boosting the growth of the solar industry. The US DoE has invested nearly $2.3 billion for intensive research and development on solar PV technology and declared a further $128 million in solar incentives. Besides the federal policies, the state-level policies such as the solar renewable energy certificates (SREC), renewable portfolio standards (RPS), the California solar initiative (CSI), feed-in tariffs (FiT), net energy metering (NEM) policies, etc. are also actively participating in promoting solar PV technology at all levels of the country. The widespread increase of microgrids, distributed generation, and energy storage systems, along with the deliberate reduction in the use of fossil fuels backed by climate concerns, are also responsible for the proliferation of the solar PV industry in the USA.

Therefore, the principal difference between the state of solar PV in the USA and other developing countries such as Bangladesh and Syria is the implementation of tax credits, policies, and mandates encouraging the development of solar PV systems. Other noteworthy differences are the economic stability, literacy rate, geographic location, land area, population density, technological prowess, etc.

7.9 Role of Solar PV in Green Energy Economy

To limit the global average temperatures to below 2 °C by the year 2100 in accordance with the Paris Agreement, it is imperative for the global communities to lower their carbon emissions and opt for a net-zero economy. The net-zero scenario refers to an equilibrium state wherein the rate of carbon capture, storage, and utilization (CCSU) will negate carbon emissions, resulting in net-zero emissions. CCSU technologies are under extensive research and development at present and are likely to reach widespread deployment within the end of this decade. At this point, we cannot but rely on the technologies we already have to lower our emissions as much as possible to protect the only home we have. Switching to clean energy systems is the most convenient option right now. Compared to other clean energy systems such as wind power, hydroelectric power, wave, and tidal power, etc., solar PV systems offer several benefits.

The solar PV system is the most easily accessible and flexible form of renewable energy for all walks of life. A PV module has no greenhouse gas emissions from energy production, is non-polluting as long as it is operational, and requires no fuel except adequate sunlight. Since PV modules are free from harmful emissions, except at their manufacturing phase, their increased deployment can substitute the conventional energy generating units, such as coal-based power plants and diesel generators. Thus, solar PV energy can play a deciding role in cutting down on greenhouse gas emissions, which is the main reason behind global warming. In the future energy mix, clean and renewable energy sources will occupy the major share, as fossil fuels have already begun to phase out. Solar PV energy, being very convenient and popular, possesses all the requirements to lead the future energy mix. The green energy economy calls for clean, non-polluting, non-depleting, and most important, inexpensive sources of energy. A solar PV system is the ideal candidate in this regard and can be the cornerstone of the green energy economy.

7.10 Conclusion

This chapter describes grid-tied PV systems with a step-by-step demonstration of the design process of a practical grid-tied PV system located in Syria. Using PVsyst and HOMER Pro software, the simulation and techno-economic analysis of the designed system are performed and discussed elaborately. In addition, the negative impacts of integrating solar PV systems into the utility grid are also described, with a focus on the scheduled outages. The chapter also contains a detailed study on the challenges and limitations of grid-tied PV systems in developing nations such as Bangladesh and Syria, in the light of comparison with a developed country such as the USA. Overall, this chapter provides a reader with three-fold benefits: first, the design process of a grid-tied PV system can be learnt; secondly, the negative consequences of grid-tied PV systems can be studied; and thirdly, an insight can be acquired about the challenges that developing nations have to overcome to make themselves self-sufficient in energy generation through the use of solar PV technology.

7.11   Exercise

1. What do you mean by the grid integration of solar energy?
2. What are the prerequisites for the grid integration of solar energy?
3. What are the steps included in the site evaluation for grid-tied PV systems?
4. Design a 2 MW grid-tied system.
5. What are the key challenges that developing countries face in the case of developing solar PV technology?
6. What are the challenges of grid-tied PV systems and their possible solutions?
7. How can solar PV technology pave the path to net-zero emissions and combat climate change?

References

1. Sierra-cascade nursery (2021). http://www.sierracascadenursery.com/home.html
2. E. Hossain, S. Petrovic, Renewable Energy Crash Course (2021)
3. M. Jaszczur, Q. Hassan, J. Teneta, E. Majewska, M. Zych, An analysis of temperature distribution in solar photovoltaic module under various environmental conditions, in *MATEC Web of Conferences*, vol. 240 (EDP Sciences, 2018), p. 04004
4. A. Diehl, Heat rise in various racking methods. https://www.cedgreentech.com/article/heat-rise-various-racking-methods
5. Solar Energy International, *Solar Electric Handbook* (Pearson, 2013)
6. Ashrae climatic design conditions 2009/2013/2017. http://ashrae-meteo.info/v2.0/
7. GROWATT, Max 50–100ktl3 lv/mv, 2021. https://www.parenasunce.com/upload/dokumenta/1580471092_MAC-30-60KTL3-LV-Datasheet.pdf
8. GROWATT, Ps-p72-(330-345w) (2021). https://www.philadelphia-solar.com/uploads/P72_1968_x_990_x35_-_June2021!!1.pdf
9. National Fire Protection Association (NFPA), *NFPA 70—National Electric Code* (2020)
10. SREDA I National Database of Renewable Energy, Solar park, 2021. http://www.renewableenergy.gov.bd/index.php?id=1&i=1
11. SYFUL ISLAM, Bangladesh outlines plan for up to 40 GW of renewables in 2041 (2021). https://www.pv-magazine.com/2020/10/20/bangladesh-outlines-plan-for-up-to-40-gw-of-renewables-in-2041/
12. S. Islam, US developer Eleris plans 2.2 GW of solar in Bangladesh (2021). https://www.pv-magazine.com/2020/12/17/us-developer-eleris-plans-2-2-gw-of-solar-in-bangladesh/
13. Power Technology, Solar power for Syria, 2021. https://www.power-technology.com/features/featuresolar-power-for-syria-stepping-away-from-the-fuels-at-the-centre-of-conflicts-5924053/
14. E. Bellini, Syria launches tenders for 63 MW of solar (2021). https://www.pv-magazine.com/2020/02/18/syria-launches-tenders-for-63-mw-of-solar/

Index

A

Absorbent glass mat (AGM), 109
Acceptor energy level, 27
Acceptor impurities, 26, 27
AC induction motor, 210–211
AC submersible pump, 216–217
Aerospace applications, 30
Air mass (AM), 78
Albedo, 75
Aluminum, 39, 150
AM, *see* Air mass (AM)
Ambient temperature, 49, 155
American Wire Gauge (AWG), 150, 151
AM index, 74
    angle z, 80
    calculating process, 79
    denotation, 78
    equation, 78
    path length, 80
    PV cells performance assessment,
        78
    radiation incident, 80
    standard value, 80
Ampere-hour (Ah), 118
Automated cooling system, 187
Azimuth angle, 83

B

Backsheet layers, 41–42
Back surface field (BSF), 38
Balance of system (BOS) components
    solar PV system, 149
    wiring process
        AC cable, 151–152
        AC distribution panel, 169
        ampacity, 150, 151
        batteries, 160–162
        cable materials, 152

        cable naming, 152–153
        cable voltage rating, 154
        copper/aluminum conductors, 157
        DC cable, 151–152
        inverter, 161–162
        polyvinyl chloride (PVC), 150
        PV modules, 160
        PV output circuit, 156
        PV source circuit, 156
        PV system, 159–160
        voltage drop, 157–161
Bandgap energy, 27
Batteries
    appropriate voltage, 119
    capacity calculation, 121–122
    cycle life, 117–118
    days of autonomy, 121
    duty cycles, 119
    electrochemical storage, 107
    life cycle, 120
    maintenance, 124–126
    maximum DoD, 119
    power generation source, 106
    PV application, 119
    PV systems
        hybrid, 108
        standalone, 107, 108
    required overall energy, 119–120
    short-term storage, 107
    sizing example, 122–125
    sizing worksheet, 124, 125
    stack sizing, 120
    temperature effect, 122
Battery bank system sizing, 197–198, 202, 206
Battery efficiency, 199
Battery energy storage systems (BESS), 107,
    204
Beam radiation, 92
Bifacial modules, 103–104